Front-Page Science

Wendy Saul, Angela Kohnen, Alan Newman, and Laura Pearce

ENGAGING TEENS IN SCIENCE LITERACY

D1517145

ENGAGING TEENS IN SCIENCE LITERACY

Wendy Saul
Angela Kohnen
Alan Newman
Laura Pearce

NSTApress

National Science Teachers Association

Arlington, Virginia

Claire Reinburg, Director
Jennifer Horak, Managing Editor
Andrew Cooke, Senior Editor
Judy Cusick, Senior Editor
Wendy Rubin, Associate Editor
Amy America, Book Acquisitions Coordinator

ART AND DESIGN
Will Thomas Jr., Director
Joseph Butera, Senior Graphic Designer, Cover and Interior Design

PRINTING AND PRODUCTION
Catherine Lorrain, Director

NATIONAL SCIENCE TEACHERS ASSOCIATION
Francis Q. Eberle, PhD, Executive Director
David Beacom, Publisher

Copyright © 2012 by the National Science Teachers Association.
All rights reserved. Printed in the United States of America.
14 13 12 11   4 3 2 1
1840 Wilson Blvd., Arlington, VA 22201
www.nsta.org/store
For customer service inquiries, please call 800-277-5300

NSTA is committed to publishing material that promotes the best in inquiry-based science education. However, conditions of actual use may vary, and the safety procedures and practices described in this book are intended to serve only as a guide. Additional precautionary measures may be required. NSTA and the authors do not warrant or represent that the procedures and practices in this book meet any safety code or standard of federal, state, or local regulations. NSTA and the authors disclaim any liability for personal injury or damage to property arising out of or relating to the use of this book, including any of the recommendations, instructions, or materials contained therein.

PERMISSIONS
Book purchasers may photocopy, print, or e-mail up to five copies of an NSTA book chapter for personal use only; this does not include display or promotional use. Elementary, middle, and high school teachers may reproduce forms, sample documents, and single NSTA book chapters needed for classroom or noncommercial, professional-development use only. E-book buyers may download files to multiple personal devices but are prohibited from posting the files to third-party servers or websites, or from passing files to non-buyers. For additional permission to photocopy or use material electronically from this NSTA Press book, please contact the Copyright Clearance Center (CCC) (*www.copyright.com*; 978-750-8400). Please access *www.nsta.org/permissions* for further information about NSTA's rights and permissions policies.

Library of Congress Cataloging-in-Publication Data
Front-page science : engaging teens in science literacy / by Wendy Saul ... [et al.].
     p. cm.
Includes bibliographical references.
ISBN 978-1-936137-14-5 — ISBN 978-1-936959-90-7 (ebook) 1. Science news. 2. Science—Study and teaching
     (Secondary) 3. Science—Study and teaching (Middle school) I. Saul, Wendy.
Q225.F76 2012
507.1'2—dc23
                    2011044161

Contents

ABOUT THE AUTHORS

vii

ACKNOWLEDGMENTS

ix

FOREWORD

By Joseph L. Polman

xi

CHAPTER 1

Science Literacy: The Big Picture

Background information on the project, including a discussion of the NRC definition of scientific literacy, our 15-years-out framework, and why science journalists became our model.

1

CHAPTER 2

Science Journalism Goes to School

A description of what a science news article is and how it functions in schools and out.

11

CHAPTER 3

Can I Do This? Frequently Asked Questions

Thoughts from teachers on how they have made SciJourn work in their schools.

23

CHAPTER 4

Science Journalism Standards

A rationale for why another set of standards is necessary, including the standards themselves. The chapter ends with a section on the ethics of journalism.

35

CHAPTER 5

Setting the Stage by Modeling

Describes how to do a read-aloud/think-aloud and the many purposes this kind of modeling can serve.

55

CHAPTER 6

"What's Your Angle?"

The qualities of a good topic, how to "pitch" stories in the classroom (slow and fast), and a list of dos and don'ts for teachers when helping students refine topics and angles.

67

Contents

CHAPTER 7

Finding and Keeping Track of Sources

Addresses standards 1 and 2 including information on searching, judging credibility, and using Diigo.

87

CHAPTER 8

Original Reporting: Interviews and Surveys

Discusses interviews and surveys from a science literacy standpoint; provides hints for how to conduct both and what to do with the results.

103

CHAPTER 9

Channeling Your Inner Science Teacher: Considering Context and Accuracy

Why should I care and other things found in the story's context, and advice for fact checking.

117

CHAPTER 10

Going the Write Way

Information and lessons on moving students from information-gathering to first drafts through multiple revisions.

133

CHAPTER 11

It's All About Revising: Moving Toward Publication

Different ways to give students feedback on their writing, including conferences, peer feedback, and other assessment strategies.

155

CHAPTER 12

Beyond Words

Products other than the article— photocaptions, infographics, Google maps, extended photo captions.

177

AFTERWORD

Final thoughts on why we find this work worth doing.

185

INDEX

189

ABOUT THE AUTHORS

In 2011 **Angela Kohnen** was a research assistant and doctoral student in the College of Education at the University of Missouri–St. Louis. For five years, she worked as a high school teacher and was one of the pilot teachers in the SciJourn project. Her research interests include writing across the curriculum and the relationship between writing and identity.

Alan Newman is a research professor in the College of Education at the University of Missouri–St. Louis. He spent 20 years as a science journalist and editor, primarily with the American Chemical Society, and holds a PhD in chemistry. On the SciJourn grant, he has served as editor for *SciJourner.org* and science journalism instructor.

Laura Pearce, a classroom teacher for over 20 years, recipient of the Presidential Award for Excellence in Teaching Science, and author of *Nurturing Inquiry: Real Science for the Elementary Classroom* (Heinemann 1999), received her M. Ed from the University of Missouri–St Louis. She works as a research associate for the SciJourn project.

Wendy Saul, PhD, serves as the Allen B. and Helen S. Shopmaker Professor of Education at the University of Missouri–St. Louis. Her work on connecting science and literacy can be found in books and articles, including the copublished NSTA-IRA volume *Crossing Borders in Literacy and Science Instruction: Perspectives on Theory and Practice*. She teaches courses in action research and heads up the professional development efforts on the SciJourn grant.

ACKNOWLEDGMENTS

In 2006 Cathy Farrar, then a teacher at Normandy High School in Missouri, called Wendy Saul, a literacy professor with an interest in science: "Could you come out and talk to my students? I want them to submit to the DuPont Challenge Science Essay Competition and they don't know how to write."

"I have a better idea for you," Saul replied. "Alan Newman, a science writer with a PhD in chemistry who for years worked as managing editor at the American Chemical Society, is in town. Maybe he'll come out there."

Newman was reluctant; he knew nothing about teaching high school. What did he have to say? Then he did what all good researchers do: he looked at the winners of the DuPont contest. Surprisingly, all the top-ranked essays seemed to follow a journalistic format. And Newman did know science journalism.

Farrar's students seemed to benefit tremendously from the help. In fact, even the weaker students who attended Newman's workshop produced better essays than some of the A students who did not attend the session. And an idea was born.

We wish to first thank Ms. Farrar, soon to become Dr. Farrar, and her students at Normandy HS. We also wish to thank the National Science Foundation and our grant officer for this project, Julio Lopez-Ferrao, for their interest and support for this project.

The three years of work we have sought to capture in this book have been undertaken by a university-based research team, two cadres of teachers, and a group from the local science center, all of whom deserve individual recognition.

The university team includes Joe Polman, PI; and co-PI's Alan Newman, Cathy Farrar and Wendy Saul, as well as Cynthia Graville-Smith, Jennifer Hope, Angela Kohnen, Laura Pearce, Nancy Singer, and Eric Turley.

Participating teachers: Patricia Baker, Tonya Barnes, Samuel Berendzen, Susan Bloor, Rebecca Bubenik, Amanda Clark, Rebecca Cook, Rose Davidson, Linda Gaither, Mary Gillis, Bridgett Gordon, Stephen Grasser, Kevin Hall, Kenna Heitman, Suzanne Hinrichs, Elizabeth Hobbs, Chris Holmes, Bobby Hughes, Patricia Hundelt, Gillian Jackson, Jennifer Jones, Mark Kasen, Rob Lamb, Sherick Powell, Michael Ruby, Alan Seder, Ken Smith, Tracie Summerville, Ellen Zerr, all need to be thanked. This book is theirs as well as ours.

At the Saint Louis Science Center a small group of students, under the aegis of Diane Miller's Youth Exploring Science program was helped immeasurably by David Hoffelmeyer, Cynthia Graville-Smith, Terris McMahan Grimes, and Rachel Mahan. That first group of students published early on in our online news magazine, *SciJourner.org*, and helped set the stage for our future research: special thanks to Adeola Adewale, Korry Busch, Jerricka Cotton, Damonte Johnson, Desiree Redus, Ariel Stavri and KiOntey Turner.

SciJourner, referred to extensively in this volume, relies heavily on three technology and design gurus and we thank them: Rosanna Cerutti, Michael Butler, and Brian Huxtable.

Keeping us all in communication and in line is Shannon Briner. She tracks the many comings and goings of a huge group, always competently and with continuing good cheer. With her own science background she also adds much to our weekly meetings.

The SciJourn group has been helped immeasurably by an advisory board. Members include: Hubert Dyasi, Kevin Leander, Julie Ann Miller, Bill Penuel, and Carol Stearns. Jay Lemke and James Paul Gee have also offered advice that was essential to our work on this project.

The authors of this volume are supported by family and friends too numerous to thank individually, but we did want to offer a special call out to Addie Driscoll, a baby literally born and bred on the project who reminds us why it is important to think about life 15 years from now, when she will be a teen.

FOREWORD

by Joseph L. Polman

I am privileged to work with the authors of this volume, along with a large and growing number of educators and researchers, as part of the Science Literacy Through Science Journalism (SciJourn) team. Our work serves as the basis of this book and for the ongoing effort to understand how the project described in this book can be useful to others inside and outside the classroom. This project has been generously funded by the National Science Foundation as part of their Discovery Research K–12 program, with the goal to create classroom practices that inform and are informed by research. The SciJourn team has asked, "What do children need to be able to do with science and technology news and information 15 years after graduation?" This has led to the formulation of standards of a different sort: ones that focus on knowledge, skills, and dispositions that students will be able to use when making personal decisions and participating in public debates years after they graduate from high school. Importantly, the science, technology, engineering, and mathematics (STEM) content that will inform those personal decisions and public debates are unpredictable and yet-to-be determined.

I come to this project as someone trained in the *learning sciences*, a field of inquiry that has contributed in important ways to the discussion of educational practices in both formal and informal learning environments. Drawing on insights from cognitive science, educational psychology, anthropology, and computer science, members of the learning sciences community have sought to build a scientific understanding of learning, and to inform the design and implementation of learning innovations (e.g., Sawyer 2006). Views of science knowing and learning born from the learning sciences have had a profoundly positive impact on education; they have directly affected reform and been taken up and elaborated on in influential reports (e.g., Bransford, Brown, and Cocking 2000; Donovan and Bransford 2005; Michaels, Shouse, and Schweingruber 2008) and have also served as the basis for science education standards (e.g., NRC 1996, 2000).

Over the last couple of decades, the learning sciences have focused primarily on how the practices of expert scientists can be used as a model and measure for what it means to "know science." Many of my colleagues and I have worked to help students "talk science," use reasoning and graphical representations, and carry out science inquiry practices, basing curricular suggestions on what practicing scientists do. We have premised our work on the notion that educated citizens in democracies should develop a strong disciplinary understanding.

The tendency to base science learning goals on an analysis of the expert scientific practices is understandable, but it has important limitations, especially if we take seriously the broader view of "science literacy" as learning that has utility years after graduation. In the learning sciences, and science education literature in general, the term *science literacy* is sometimes used synonymously with the ability to carry out the firsthand inquiry practices of expert scientists—see, for example, my own and colleagues' work on educating "little scientists" (O'Neill and Polman 2004). With this kind of focus, the literacy in *science literacy* may become lost. But a science literacy that includes sense making, reading, writing, and communicating about contemporary science topics as they relate to everyday life

and policy making is obviously important to life and citizenship. That is the kind of science literacy toward which SciJourn aims.

The model in this case is not the lab scientist, but rather the science journalist. Once we realized that the science journalist was an excellent expert model for the kind of science literacy we were targeting, we used techniques from cognitive science to analyze expert practices of science journalism, namely clinical interviews, think-aloud protocols, and task analysis (e.g., Ericsson and Simon 1993). Those aspects of expert science journalists' practices that aligned most strongly with the science learning goal for 15 years after graduation were elaborated to inform the SciJourn standards (Chapter 4).

Keeping in mind what people making personal decisions and participating in public debates can learn from good science journalists only gets us so far, however. Learning sciences research has taught us a good deal about the cognitive, social, material, and cultural aspects of organizing learning environments, and the SciJourn approach that is described in this book takes those into account as well.

Research has shown that for learning and development to take place, experiences must have an authentic meaning for participants so that they remain engaged. And learners must also have effective scaffolding that builds on their prior knowledge and serves the tasks at hand. Throughout this book, you will read examples of science and technology news stories developed by and reported on by young people and that relate to the students' own prior experiences and interests, whether it be the health condition of a neighbor or family member, participation in a sport or leisure time activity, or a decision to be made in their local community. By engaging teens as citizen science journalists in an educationally supportive environment, young people build the knowledge, skills, and dispositions they need to answer future questions about their hobbies, their jobs, their communities, and their own or a loved one's health.

Scaffolds, activities, and products such as science news stories are important because our current theories and recent research in the learning sciences has reinforced the notion that we must understand learning and development as occurring "beyond the skin" of individuals, and we should view cognition and action as inherently social and cultural. Thinking and acting involve the cultural tools that human beings inherit and are inextricably bound to the situations and contexts within which we act. This is one reason why the strong separation many students feel between school and the "real world" is a problem. It is notoriously difficult for students to use much of the knowledge gained in schools precisely because the social and cultural arrangement of activities in schools is too often unrelated in both the students' and the teachers' minds to future activities. Learning takes place through the social and cultural practices and processes that occur within the context of the many communities each of us encounter in our lives.

The SciJourn model conceived, tested, and refined in this book lives within a loosely connected network of learning communities in school classrooms and in a youth development program. Within this network, we have sought to create hybrid spaces that support learning and development, but that also remain connected to the real worlds of science and technology research and development, as well as the local communities in which these sites are embedded. We envision the young participants in each of these learning communities as travelling along trajectories—from their past lives, through the pathways that we create together, and toward futures that they will forge. We know that all of these learners have the potential to become more scientifically literate adults who are committed to and capable of using science and technology information to enrich their lives, and the lives of those with whom they interact. Our goal is that they become the kind of people who are not afraid of science; who recognize how science is relevant to their lives, can find information and make sense of it using multiple and

diverse sources, understanding what each has to offer, and placing that information into the context of prior research and applications for society.

I hope that this volume inspires you to create new and unique hybrid spaces that make use of and expand on the ideas you find here.

References

Bransford, J. D., A. L. Brown, and R. R. Cocking. 2000. *How people learn: Brain, mind, experience, and school*. Washington, DC: National Academies Press.

Donovan, S., and J. Bransford. 2005. *How students learn: Science in the classroom*. Washington, DC: National Academies Press.

Ericsson, K. A., and H. A. Simon. 1993 *Protocol analysis: Verbal reports as data*. Cambridge, MA: MIT Press.

Micheals, S., A. W. Shouse, and H. A. Schweingruber. 2008. *Ready, set, science! Putting research to work in K–8 science classrooms*. Washington, DC: National Academies Press.

National Research Council (NRC). 1996. *National science education standards*. Washington, DC: National Academies Press.

National Research Council (NRC). 2000. *Inquiry and the national science education standards: A guide for teaching and learning*. Washington, DC: National Academies Press.

O'Neill, D. K., and J. L. Polman. 2004. Why educate "little scientists?" Examining the potential of practice-based scientific literacy. *Journal of Research in Science Teaching*, 41(3): 234–266.

Sawyer, R. K., ed. 2006. *Cambridge handbook of the learning sciences*. New York: Cambridge University Press.

CHAPTER 1

SCIENCE LITERACY: THE BIG PICTURE

Teachers—and schools in general—face a daunting challenge as they seek to prepare students for a future that cannot be predicted with any degree of certainty. Although the difficulties of helping to develop a curriculum that addresses future needs are evident in the humanities, the arts and mathematics, for those of us in science, the problem sits like the proverbial elephant in the classroom, staring at us right here, right now.

A few examples: The U.S. Preventative Services Task Force issued new guidelines for mammograms in 2009. What does this mean for someone with a family history of breast cancer? Congress periodically votes on a piece of legislation called the Farm Bill. What does its current iteration mean for the safety of supermarket eggs? Understanding how the latest science affects real people—patients, consumers, voters, and taxpayers—is at the heart of science literacy.

Here is the question we seek to address in this book: What might we teach students now that will help them make sense of their world 15 years after graduation? While we strongly believe that it is important for young people to learn underlying scientific concepts in school, we also realize that science changes faster than curriculum or textbooks can be updated. DNA will still be the basis of inheritance and matter will still be made up of atoms and molecules, but the treatment for type 2 diabetes or Crohn's disease will likely change in 15 years and maybe quantum computers or salt-free diets will be the latest craze. To understand and engage with scientific, medical, and technological issues 15 years after graduation, students will need skills that both allow them to continue learning about science concepts and enable them to find, make sense of, and assess the information they encounter.

Although all of us are concerned about the end-of-course evaluations and national standards, we know that the mark of our success as teachers will not be scores on a test administered at the end of high school. Rather, our success or failure as educators has to do with how students will do in life—as parents, as voters, as members of their community. The question finally comes down to this: What can we do in school to promote *life* success?

Of course, we are not the first group to recognize and seek to address the issue of science literacy. The National Science Education Standards, the blue-ribbon panel from the National Research Council (NRC), developed a multipart definition of *scientific literacy*. It asserts that a scientifically literate person can

> ask, find, or determine answers to questions derived from curiosity about everyday experiences. It means that a person has the ability to describe, explain, and predict natural phenomena. Scientific literacy entails being able to read with understanding articles about science in the popular press and to engage in social conversation about the validity of the conclusions. (NRC 1996, p. 22)

Clearly, there is plenty of justification for addressing the connection between science and issues of personal relevance. But the question remains: Where in the curriculum do we invite such queries and such research? What resources do our school systems provide to help students answer those questions? How do we assess their ability to make sense of information? With a single textbook, limited access to the internet, and library budgets on hold, the possibilities for question-posing and answer-seeking indeed seem limited. Moreover, we are faced with a trifecta of stumbling blocks—quickly evolving science information, fundamentally static textbooks, and declining school budgets that do not support other print resources.

Other attempts to address issues that fall under the heading "science literacy" are surely evident. Many teachers ask students to write reports on contemporary topics, such as global warming, stem cells, or recycling, and articles from the popular press can be used as citations. If students work on these reports in cooperative groups, perhaps some conversation even occurs. But in truth, the important skill of being able to read contemporary articles of personal import with understanding—a skill surely needed throughout an individual's life—becomes a kind of add-on. In our reading courses and through mandated testing programs, do we do enough to teach the difference between basic comprehension and critical analysis? Too often, we are satisfied if a student has been able to track down and rephrase an article about a particular topic. Rarely in the classroom do we teach students when and how to be skeptical, where to find credible information, and when and how to seek additional support for an assertion.

The NRC wants students to do more than simply collect information:

> Scientific literacy implies that a person can identify scientific issues underlying national and local decisions and express positions that are scientifically and technologically informed. A literate citizen should be able to evaluate the quality of scientific information on the basis of its source and the methods used to generate it. (1996, p. 22)

In essence, what this statement says is that basic comprehension is not enough. The NRC, and we, want students to learn enough about a field so that they are not fooled by, or at least remain suspicious of, certain claims. In addition, educated, scientifically literate citizens should recognize the difference between various publications—peer-reviewed journals, fact-checked newspaper and magazine accounts, TV and radio talk show hosts in the business of "edutainment," or the wild west of online forums. The definition proffered by the NRC clearly asks teachers to move scientific literacy beyond the classroom walls. To be scientifically literate is to be able to understand issues that are important to local communities, issues that may vary from school district to school district, and issues that cannot always be identified by textbook makers or curriculum coordinators years in advance.

In its definition of science literacy, the NRC does, in fact, turn to the issue of argument: "Scientific literacy also implies the capacity to pose and evaluate arguments based on evidence and to apply conclusions from such arguments appropriately" (1996, p. 22).

Teachers have typically been asked to scaffold students' attempts at written argumentation by creating a formula: A paper is assigned that requires something like a clear thesis statement with at least three supporting points, each backed with at least two supporting details, followed by a clear conclusion. To help teachers respond to and grade student papers, the educational establishment offers scoring rubrics. In our attempts at clarity and support, however, we may be scaffolding their work to the point of removing opportunities for thoughtful decision-making. After all, teaching nonformulaic writing is more than teaching compositional form. Good writing is essentially about understanding, critical thinking, and clarity. And teaching students to think—and helping students learn to think—is always messy work. Relatively simple forms like the five-paragraph essay, a form teachers hope students might be able to grasp in time for the next required test, are, frankly, contrived and inauthentic. Students may be able to turn in a paper that conforms and gets an acceptable grade, but if we are looking for an authentic performance that demonstrates science literacy, this is not it.

Instead of beginning by identifying a product (a five-paragraph essay, an editorial, a newspaper article) and promoting it as the key to science literacy, we began instead by asking: Who in the real world (as opposed to the school world) do we view as scientifically literate and what can we learn from them that might be transferred in whole or part to the classroom? Our goal is to look toward authentic examples rather than to contrive an answer based on school conventions. The likely suspects, of course, were highly educated scientists. They know a lot of science, so surely they are scientifically literate, we assumed. As part of a project supported by the National Science Foundation (NSF), we began by spending time with these folks, trying to trace what these scientifically literate individuals did as they read, wrote, and talked.

What we learned is that even people who are truly expert in their field have all sorts of "holes" in their science education. This seemed like altogether good news if you recall that our goal was not to find people who know today's science, but rather those who are equipped to deal with science of the future, science that is not yet available or known to the "consumer." With that in mind, we asked scientists to read and think aloud about a science article that was clearly out of their area of expertise. One PhD biologist, for example, reading an astronomy article in a popular science magazine, turned to the interviewer in the middle of the article and said, "You know, I have no idea what I am reading!"

> Interviewer: "So, what could you do at this point if you wanted to learn more?"
>
> Biologist: "I'd probably ask a friend, if I cared, and if I didn't, I'd put the article down and not worry about it."

Let's look at this exchange. The first thing to notice is that the scientist did not get upset or suggest to the interviewer (or anyone else) that she is "bad at science" because she could not understand the article. This fits nicely with what Jay Lemke, author of *Talking Science* and a scholar well-versed in physics, linguistics, and cognition, wrote in an exchange on the topic. Lemke thinks of science literacy as "the embodiment of an attitude: a sense of self-confidence about one's ability to cope with ideas, concepts, information, or practices that derive in whole or in part from research in the natural sciences" (2009).

Alan Newman, a PhD chemist and our science journalist on the NSF project, talks about having a similar attitude, saying he's not afraid of science: "I don't know much about geology, for instance, but I am not afraid to find out. I can connect what I am reading to something I already know about and start there."

The second point to note about our scientist-reader is that she's ready to ask a friend. Again, Lemke has something to say about this: "What do people do when they want to understand something and come up against an element that is unfamiliar, though perhaps in the context of many other elements which are

familiar? They ask someone." This simple learning strategy is the basis of the view of science literacy proposed by Wolf-Michael Roth and others that science literacy is an attribute of a community, not of an individual (Roth and Lee 2004).

Perhaps the goal of educating people to become scientifically literate is a bit off. Perhaps no individual can truly be scientifically literate, but he or she can and should have access to, and be seen as a part of, a community that is scientifically literate. Or, as physicist Morris Shamos argued, we should learn to "use" experts (1995). With the help of that expert, distributed community, an individual can make informed decisions about issues of personal or civic importance. In this view, knowledge is distributed within the community. For those of us whose days are spent fixing cars or typing away in an office, there should be a clear sense that we have both the confidence to seek information and the knowledge and ability to ask for credible help when our own resources are wanting.

We know that working scientists, physicians, and engineers must be a part of this distributed community, but they are often in their own separate intellectual neighborhoods, talking mostly to other specialists. If working and well-educated scientists are not our model for the scientifically literate individual because their training sets an impossible standard, who might we turn to instead as embodying the qualities we seek in a scientifically literate public? Is there a group working in the world who might serve as a model for students with broad interests—or for their teachers trying to prepare students for this uncertain future?

Enter Science Journalism

Traditionally, and ideally, journalists have served as the public's educator and interpreter of what is new and controversial in science and in other fields as well. We count on them to do the research and to come up with as close to an unbiased description of topics of current interest as possible. In fact, newspaper and magazine editors have historically served as gatekeepers, attending to issues of credibility, which is why most papers don't print the equivalent of "Aliens Made Me Pregnant!" stories.

Because science journalists are tasked with finding and evaluating information the same way that scientifically literate people need to find and evaluate information, teaching students the process of science journalism helps them find credible information and make informed decisions years down the road. Unfortunately, the role of journalist-gatekeeper is disappearing with the rise of the internet (the wild west of ideas—democratic, but unvetted) and the decline of traditional print journalism. For this reason it may be even more important that students develop a citizens' sense of themselves as information critics.

There are other reasons why science journalists serve as a powerful model for those seeking to become scientifically literate. Many science journalists don't have a science background, and the ones who have studied science often write outside their area of expertise. What they do know, however, is how to get their story. They are the trackers and hunters of science information. To that end, good reporters

- formulate a question of current interest to their readership;

- investigate the question by finding and querying multiple, credible sources;

- digest the information—including various sides of controversies, whether the science is embryonic or a consensus view—and the relevant technical information; and

- summarize the information, ideally without bias, in a way that the general public can understand.

In short, good reporters become literate, but not necessarily expert, on a scientific topic. And they learn how to negotiate what might be called the landscape of science information, while at the same time becoming part of it.

Reporting—making calls, interviewing experts, reading articles, taking lots of notes, and analyzing the collected information—makes up the bulk of the journalists' work. Writing is the creation of a coherent article or story from these mounds of information. Most of this book's lessons focus on reporting, because this skill improves students' ability to understand and make decisions using scientific information. We call this project "SciJourn," and the online and print publications that seek to engage teens in science literacy though science journalism are *SciJourner.org* and *SciJourner*, respectively. The project is purposefully designed to invite variations and adaptations to suit the needs of schools and other informal educational settings.

In our studies of the SciJourn approach in high schools, we have found that teens don't lack for good ideas to write about—for instance, the health risks of tattoos, why a tennis ball can "pops" when it opens, how can schools be "greener," is artificial or natural grass better for sports fields, and the nature of dozens of common and uncommon diseases. But helping students move from a general topic of interest to the reporter's approach—formulating a good angle or approach to a topic, finding and querying sources, assessing credibility, digesting the most recent information, recognizing what they do and do not understand, and summarizing without bias—is what this book, and what science literacy, is finally about.

Health Literacy

Although our focus is on science literacy in general, health literacy has emerged as an important subset of concern. Unlike science literacy, health literacy is mostly defined as an adult issue: grown-ups unable to navigate the health care system, read the instructions that come with their prescribed drugs, or follow the instructions given by their doctor (Nielsen-Bohlman, Panzer, and Kindig 2004). The ironic aspect is that for most Americans the last formal health class they encounter is in high school. And the topics typically found in these health classes—smoking cessation, HIV, STDs, unwanted pregnancy, injury prevention, human sexuality, and so on—are rooted in teenage concerns (CDC 2006).

Because SciJourn uses a forward-looking model for science literacy, it anticipates those adult health literacy problems. Finding up-to-date and accurate information from multiple credible sources, learning to interview expert sources, and critically digesting technical information are the skills needed to understand health issues.

Moreover, health literacy resonates with teens also. By far the largest number of articles submitted to *SciJourn.org* for publication focus on health issues. Some of the topics are the same ones found in high school health classes; healthy eating is particularly popular. However, many deal with diseases that get short shrift in high school, such as anal cancer, Down syndrome, or congestive heart failure, concerns born both from personal experience and from media stories on favorite stars. Other students tackle rare diseases that would even challenge many doctors, such as the autoimmune disease pemphigus or the fatal genetic disease Potter's Syndrome. We have even published a story on Rare Disease Day, which was established to raise awareness of these unusual maladies (first observed on February 29, 2008, itself a "rare day"). No standard curriculum would even consider such content, but for the teen whose neighbor's child died or the one whose aunt suffers from such a malady, the concern is palpable.

Figure 1.1

A Recent Print Edition of *SciJourner* Featuring Student-Written Stories

Regional Coverage   World Exclusives

Teens Engaging Science Through Journalism

Volume 3 • Issue 1   APRIL · 2011

Recycle Yourself

ROSE RITTER
Fort Zumwalt North High School

There is an easy way to recycle yourself—fill out the back of your driver's license.

Each year when they obtain their driving permits, thousands of teenagers are asked if they will consent to become an organ donor. What teenage drivers do not realize is a simple "yes" to becoming an organ donor could save the lives of at least eight other people, according to health-first.org, an organization of Florida-based hospitals. In a recent study, donatelife.net, an educational site created to inform Americans about organ donation, concluded, "90% of Americans support donation, but only 30% know the essential steps to take to be a donor."

"More than 19,000 organ transplants are performed in the United States every year," reports health-first.org. However, approximately 106 people are added each day to the organ waiting list, and 18 die every day waiting for an organ, as reported by the University of Pittsburgh Medical Center. According to donatelife.net, there are over 100,000 people currently waiting for a transplant.

Kidney failure, heart disease, diabetes, and cirrhosis of the liver (a deadly liver disease) are just a few conditions that could require an organ transplant, according to the Organ Procurement and Transplantation Network (OPTN).

Robert Truog, Professor of Medical Ethics at Harvard Medical School, told the *New England Journal of Medicine* in 2008, "the growing transplant waiting lists obligate us to strive to increase the supply of transplantable organs."

For Missouri residents, becoming a donor is simple: Join through any driver license office or register online at missouriorgandonor.com. It is vital for donors to inform their family because they make the final decision about organ donation.

Practically anyone can be an organ donor. Most religions support the act of donating organs and view it as a selfless operation of charity and love, according to organdonor.gov. No one is too old or too young to donate, because it is not age but the condition of the organs that is most important, according to organdonor.gov. Only those with the Human Immunodeficiency Virus, cancer, or disease-causing bacteria or viruses in their blood are barred from donation. Another exception are potential donors who suffer a cardiac death, in which the heart stops beating, because the organs deteriorate too quickly due to the lack of oxygen, according to health-first.org.

There are two types of organ donations: nonliving and living. In nonliving donations, hospitals contact the Organ Procurement Organization (O.P.O.) upon the donor's death, according to OPTN. The O.P.O. then evaluates potential donations and finds a corresponding patient. The organs are then extracted through a small incision and are kept in a cooler during transportation, according to organdonor.gov and OPTN. Organs can only be stored between 6 to 72 hours.

Blood is the most common type of living donation since it is continuously being replaced in the body. Blood can be donated every 56 days, according to mayoclinic.org. Parts of the liver, pancreas, lung, and the entire kidney can be transplanted from a living donor, according to transplantliving.org. Living organ donations have become more frequent in the U.S. during the last two decades, with 5,452 in 2010 compared to 2,123 in 1990, according to OPTN, even though these donors must deal with recovering from elective surgery.

Most families find that organ donations from a loved one help them go through the sorrow of death, according to health-first.org. They are comforted by the thought of a positive benefit to their loved one's passing.

Luc Noel, Coordinator of Clinical Procedures at the World Health Organization, states, "Donation of organs after death, when you don't need them any more, is a civic gesture."

With the lack of human organs, some European countries are requiring all citizens to be organ donors unless the government is notified otherwise, according to www.anatomicalgiftact.org, a website that facilitates organ donation.

Ever Heard of Alternative Medicines?

A.J. PARK
Rockwood Summit High School

When traditional medicines failed, the author turned to cupping—an alternative medicine. Was that a good idea?

Ever since the beginning of middle school, and perhaps even earlier, I've suffered from chronic, energy-draining headaches that would sometimes leave me dizzy and light-headed. Even through what seemed like endless bottles of over-the-counter and prescribed medicines, my headaches could not be alleviated. Seeing my struggles, my grandmother suggested that I go see a friend who was an herbalist.

What I got out of the visit was this: Cupping, which is usually described as a procedure in which a rounded glass cup is warmed and placed upside down over an area of the body, creating a suction that holds the cup to the skin. Cupping increases the flow of blood, according to the National Cancer Institute

(NCI) and the National Center for Complementary and Alternative Medicine (NCCAM).

Weekly use of cupping didn't completely rid me of my headaches. I still have minor headaches, but they weren't and aren't as high in the pain scale as the ones I used to have. However, cupping wasn't the only alternative medicine I could have turned to.

Alternative medicines are "practices used instead of standard treatments that generally aren't recognized by the medical community as standard or conventional medical approaches," according to the NCI. Examples are acupuncture, yoga, eating healthy, and supplements.

Alternative medicines are now more commonly recognized as complementary and alternative medicine (CAM), "a group of diverse medical and health care systems, practices, and products that are not generally considered part of conventional medicine," according to NCCAM. Formed in 1998 by the National Institutes of Health, NCCAM is the "lead agency for scientific research on the diverse medical and health care systems, practices, and products that are not generally considered part of conventional medicine."

I'm not the only one benefiting from CAM; it is becoming a big help in medical fields. For example, researchers at the Oregon Health and Science University reported in 2010 that yoga exercises can help with fibromyalgia.

Nowadays, CAM is being paired up with conventional medicines, according to NCCAM. An example would be pairing various Chinese herbs with proven allergy treatments.

The pairing approach was studied by the American College of Allergy, Asthma and Immunology (ACAAI) in Phoenix. Allergist William Silvers from ACAAI concluded that "the use of substances found in nature, such as herbs, foods and vitamins, can be helpful in treating various allergies when combined with traditional therapies." He also said "it is important that patients consult their allergist before adding [CAMs] to their treatment plan."

It is important to ask educated medical professionals when using CAM with and without conventional medicines, say experts. *Discovery Health*, which is the Discovery TV health channel, says some alternative medicines are put out to the public with misleading information about the product. It also says that not all side effects are known, so some of them could be potentially harmful. Finally, they warn that people sometimes end up relying solely on CAM and stop using conventional medicines, which could cause death.

MOBEEN MIAN
Schools are huge producers of waste. Is your school eco-friendly?

How Green Is Your School?

MOBEEN MIAN
Francis Howell High School

Ever wondered how much trash comes out of a school? Schools are a huge producer of waste. Can schools be more eco-friendly?

According to *Earth Share*, an environmental group dedicated to promoting environmental education, one student will produce on average 67 pounds of waste in one month by bringing lunch to school. This adds up to 18,000 pounds a year. The majority of the wastes are paper materials and food waste, according to *Recycle Now*.

If schools operate in a more environmentally conscious way, not only does it create a much cleaner environment and reduce pollution, but it also lowers the school's cost of operating. For example, research

Continued on page 3

The point is that of all the science literacy skills teens need, health literacy is perhaps the most neglected and the most urgent—both personally and, given the tremendous costs, publically. Yet, when we talk to teens about credible sources, we find that few have heard of the Centers for Disease Control or the National Institutes of Health. Many never thought to look to nonprofits such as the American Lung Association, Planned Parenthood, or the American Heart Association for medical information. The idea of questioning an adult, especially a health care professional, is scary. And a surprising number of teens are willing to believe unsupported medical or health advice from friends, an anonymous 16-year-old writing a blog post on the internet, or from a product-promoting huckster or television personality.

About This Book

Though this book may look like a guide for teaching science journalism, it really asks that readers view science journalism as a means of helping students become scientifically literate or, more correctly, to become better consumers of and contributors to a scientifically literate community.

The book itself is basically divided into three parts and readers are encouraged to skip around, gathering ideas on a "need to know" basis. Chapters 1–4 offer background information and a rationale for using science journalism. We describe what we mean by *science literacy* (Chapter 1), deconstruct the science news story (Chapter 2), and cite and answer frequently asked questions about the nitty-gritty of working though a year with SciJourn (Chapter 3). Chapter 4 identifies and explains our standards and is rich with student examples.

The second part of the book offers concrete advice about how to teach the kind of science literacy we argue for in the first section. We describe read-alouds/think-alouds, finding an angle to write about, search strategies and the technology to support searching (and not losing your notes!), and interviewing. Remember, in this book we are not debating which subjects, concepts, or grade-level expectations are most important. Our goal is to help create curricula that teach students to deal with information that *is* specific but that we cannot specify in advance. It is born from students' interests and concerns. Some suggest that this is, in fact, a sea change that enables us to think about the purposes of school a little differently.

The remaining three chapters are about the process of putting together and writing a news story. It offers teachers ways to get students going, how to help when they are stalled, and how to respond to their drafts and revisions. We have come to view the science news article as a display of a student's scientific literacy that enables a teacher to see and talk about science literacy goals. We also offer advice to teachers on how to support both the more- and less-able writer. An afterword offers a few ideas to ponder as you move ahead with your own version of SciJourn.

This book is supported by a free website called *Teach4SciJourn.org*. Each chapter is attached to teacher suggestions and offers a place for comments

and queries. Lesson plans are offered in downloadable form and a section on "news" is designed to keep you up-to-date with SciJourn developments.

We view the SciJourn project as its own distributed community. Welcome.

References

Centers for Disease Control and Prevention (CDC). 2006. *School health policies and programs study.* Available online at *www.cdc.gov/HealthyYouth/shpps/2006/ factsheets/pdf/ FS_Overview_SHPPS2006.pdf*

Lemke, J. Personal communication to Wendy Saul. February 2, 2009.

National Research Council (NRC). 1996. *National science education standards.* Washington, DC: National Academies Press.

Nielsen-Bohlman, L., A. M. Panzer, and D. A. Kindig, eds. 2004. *Health literacy: A prescription to end confusion.* Washington, DC: National Academies Press.

Roth, W.-M., and S. Lee. 2004. Science education as/for participation in the community. *Science Education* 88 (2): 263–291.

Shamos, M. 1995. *The myth of scientific literacy.* New Brunswick, NJ: Rutgers University Press.

CHAPTER 2

SCIENCE JOURNALISM GOES TO SCHOOL

W here in school have we taught students to thoughtfully gather and assess the information needed to write and think well? The curriculum in science classes is packed to the rafters. Science teachers battle the calendar and the clock to fit in everything they are required to cover. Most hope that students learn to research and read critically in English. However, English teachers, in order to concentrate on the writing process, often rush through the research process, either by selecting topics or websites for their students in advance or not attending to the questionable internet sources being used. Students, anxious to get their work done, begin writing before they have fully comprehended the issues they are writing about. Formulaic writing (i.e., a concentration on genre or form) can contribute to this problem. Teaching to prepare students for future tests or academic tasks usually has little relationship to personalized questions or authentic work.

The approach we explain and advocate in this volume may be one of several that use authentic, contemporary tasks as a way of building science literacy into the curriculum, but it is an approach that seems to work. For the past few years, with support from NSF, we have engaged students in thinking and acting like science journalists. Through SciJourn, teens build their skills as scientifically literate individuals and their efforts are realized in both an online magazine, *SciJourner.org*, and *SciJourner*—our print publication.

The SciJourn project's goal is to make interest in science more concrete and contemporary, more applied and relevant, more a part of students' everyday lives. This is not to suggest that schools should ignore basic science information and principles or that there is no place for standards-identified content. Rather, our hope is that this project contributes to the educational landscape by helping students view science as a tool and a way of thinking that has utility—personal utility. We do what we can to help them figure out how to live a life characterized by thoughtful decision making based on credible information. At present, our students may be able to decode and perhaps even comprehend informational texts, but they have not learned how to find, use, analyze, and evaluate information, nor have they learned how to share the science they care about with others.

As educators, we have probably done little to model the kind of thinking we seek to promote among students. How often in our class notes or lectures do we attribute sources or say why we find a particular theory credible?

Perhaps the most prominent player in this sad story is the textbook. In an attempt to transmit information most efficiently to young people, we have purchased and relied upon textbooks. Information in these tomes is packaged and information is simply asserted. The textbook is *the* source, the site of wisdom. Students are not invited to comment on nor critique the content. Moreover, virtually no information is attributed. It comes to students as complete, uncontested, and shrink-wrapped. No notion of who is an expert and why we believe one expert as opposed to another is included in the typical textbook. Authors of textbooks, especially those aimed at novices, are typically not grounded in a real-world context; they are removed from the real science problems students are likely to encounter. Even when textbook authors seek to ground concepts with concrete examples, the examples are often artificial, designed to remain uncontroversial, nonspecific, and frankly impersonal. SciJourn provides for the opposite; examples are authentic, sometimes controversial, highly specific, and definitely personal.

The textbook approach is riddled with other problems we seek to address through SciJourn. Knowledge appears to be generated from a single source rather than viewed as distributed among a community of experts and non-experts, generated from extensive processes of trial and error. Moreover, the textbook leaves no obvious place for students to enter the discourse of science. We often begin a SciJourn professional development session by asking participants—and you can do this with students too—to describe the difference between a newspaper article and a textbook. (For a fully elaborated version

Figure 2.1

The *SciJourner.org* Homepage

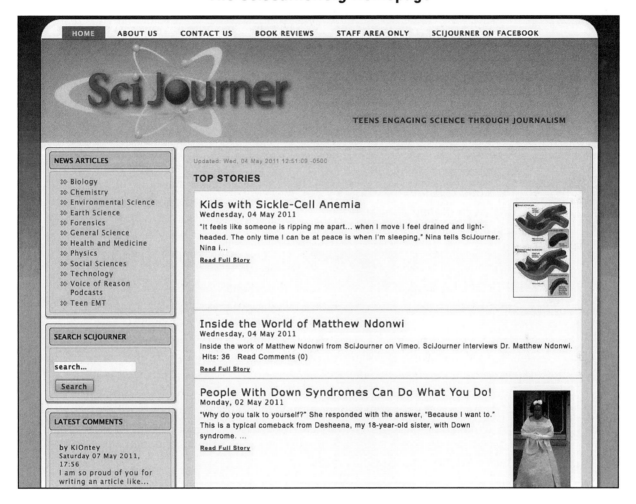

of this lesson, see our accompanying website, *Teach4SciJourn.org*.) They may note, for instance, that news articles in general (and student-written *SciJourner* articles in particular) identify various experts and quote these experts "in conversation" with others in the community.

Ideas regarding expertise are another clearly noted difference. For example, when looking at waste in the cafeteria, the school lunchroom staff is recognized as one important source of information along with an expert on waste management. When doing an article on teen pregnancy, talking to teen mothers makes sense and gives the article's author concrete issues to address as she provides a context for her story. *SciJourner* authors have been known to poll their classmates on attitudes and behaviors and garner large numbers of responses. They

"As an English teacher I am haunted by my failure to teach the basic tasks of accessing, applying, and assessing information. When students don't learn to connect books, articles, and internet information to real problems, they end up unprepared to make real life decisions. Real life learning just doesn't fit into the school day as presently conceived."
— SciJourn teacher

also mine government agencies and nonprofits for relevant data and "translate" technical terms into standard English. The student authors turn absolute numbers into analogies, comparing, for instance, the size of a spore to the thickness of a hair or the dimensions of an iceberg to the size of a state. In so doing, the writers make numbers more understandable. Moreover, we encourage readers to raise questions in the teacher-led discussions and leave comments on *SciJourner. org*. Our overarching goal is to foster questioning, analysis, and critique.

This chapter provides a detailed description of what we mean by a science news article and gives the reader a language to describe how such articles work. The following chapter, a list of frequently asked questions and answers, is designed to help teachers who agree with the principles set out here but are nervous about wading into the waters of science news writing; it offers concrete advice to address the worries of those reluctant to take a dip. If you are a teacher who is more worried than curious about the work we are promoting, by all means, go straight to Chapter 3, and if we have convinced you that SciJourn is possible to enact in your classroom, rejoin us as we describe the nuts and bolts of science journalism!

Science Journalism

What exactly is a science news article? To understand that, it is important to differentiate journalistic writing from other forms of expression. The job of the journalist is to quickly impart important information—what is new and significant—to a broad readership using clear, easy-to-understand language. Because the information is new, it may be controversial or, in the long run, wrong. To deal with that, the journalist relies on comments from multiple, credible sources, some of whom are not associated with the work. Much of the work of a journalist is just getting to the right sources and gathering some comments. Asking the right questions of the right people is the ultimate goal.

Science journalists have an additional skill—they understand at least some of the underlying science and they appreciate the nature of the scientific enterprise. Thus, the political reporter pits democrat against republican or conservative against liberal and their job is done. But the good science reporter does not "balance" a finding by an evolutionary biologist against a creationist, but rather asks the evolutionary biology expert at Cornell University to comment on similar work taking place at the University of Illinois. Moreover, the good writer does not cherry-pick a single study and hype it as something proven— the bad or good food/drug/technology-of-the-week problem—but tries to put the science into perspective. In short, the science journalist acts as a credibility gatekeeper (Blum, Knudson, and Henig 2006; Kovach and Rosenstiel 2007).

Aiding the journalist is the editor, whose job is ideally to serve as a proxy for the reader. The editor approves the story because the publication's readers will be interested. The editor also content edits (not copyedits) the story to ensure that it makes sense, seems credible, tells the story effectively, and answers the readers' questions. Both the author and editor are in effect storytellers and both work as gatekeepers on what the reader finally sees.

All of this may seem like a tall order for teens and teachers, but what we have found is that many high school students do have science stories—really good stories—and only need to be guided to credible sources and challenged to articulate the relevant details in order to create something informative. And, if the teacher is willing to be a colearner, educators can be effective guides and coaches for their students.

Characterizing a News Story: The Inverted Triangle

Journalistic articles have a structure, but we can't emphasize enough that treating the structure described next as a box to fill is an error. Becoming overly focused on the news article's form creates the same problem that we see with the ubiquitous five-paragraph essay we discussed in Chapter 1. Instead, consider the journalistic article an ideal and a way to understand news stories written by professional journalists. Keep in mind that the journalist is first a credible, honest storyteller.

Look at the diagram of an inverted triangle (Figure 2.2, p. 16). Journalists often describe their articles in this way, making sure that the most important information is in the forefront. In other words, the very beginning contains the key parts of the story. Reading only the first few paragraphs should present enough information so that the reader does not even need to read the full story! Journalists understand that. In fact, in the days of print, stories that didn't fit their allocated length would be summarily cut and their "unimportant" last sentences tossed out.

Figure 2.2

The Inverted Triangle

Typically, journalistic articles are structured in roughly this order. The wider the triangle (and the larger the type font), the more important the information is.

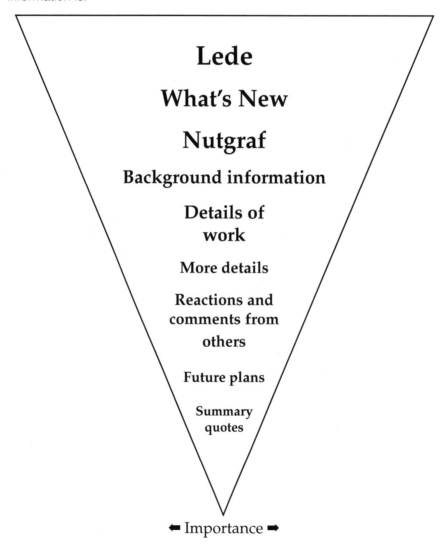

Lede

The *lede* (rhymes with seed) is the beginning sentence or paragraph; it is designed to capture a reader's interest, like a blinking neon sign. Check out these attention-grabbing ledes created by SciJourn students:

- "Have you ever purchased fresh produce only to discover that it has rotted a few days later?" (Haas 2011).

- "RING! The school dismissal bell sounds, forcing you out of your nap and back into consciousness" (Macheca 2011).

- "In April, 2009, my grandfather, Christopher Dalton, age 56, made 'the hardest decision' of his life" (Vehige 2011).

Ledes can be as short as a single word—"RING," for example—a paragraph-long story, or if it is serious enough, the news itself, as in this example: "Chlamydia was the most frequently reported bacterial sexually transmitted disease in the City of St. Louis in 2008, according to the CDC" (Emrick 2010).

Nutgraf

That first paragraph includes a succinct statement of what is new and important and is called the nutgraf (for "nutshell paragraph"). It previews what the article talks about in detail later on. For example:

- "'Every 3 seconds an iPad is sold,' says Apple's chief executive, Steve Jobs. How can one device sell so well? The design of the user experience may be responsible" (Busch 2010).

- "According to John Timmons, President of the Three Rivers Beekeepers in St. Peters, MO, beekeepers in Missouri are reporting that their bees are less affected by colony collapse disease. What makes Missouri different than other states?" (Smith 2011).

- "When you drink energy or sports drinks, do not brush your teeth immediately after drinking, say New York University dental researchers. The acid from these drinks softens tooth enamel and if you brush your teeth immediately, you will brush your enamel away" (Lee 2010).

Notice that attribution (see definition on p. 19) is an important part of the nutgraf. Getting that specificity of who is saying what immediately establishes the credibility of that story. Also notice that the author sets up a "problem": Why does the iPad sell so well? Or has a different spin on a typical story, what journalists call an angle: Why is Missouri different than other states? The reader anticipates a story that explores that problem or angle. Hopefully, the reader is also thinking, "Hey, why is that?" and is hooked.

Contextualization/Background

The reader can't appreciate the story without putting the issues into context. Why is this important? How many people are affected or how many of these devices are in use? What came before and what are its limitations? What laws or rules are in effect? Examples include

- "In a 2009 report by Raymond Cloyd, Associate Professor for the Department of Entomology Extension at Kansas State University, the loss of honeybees in the United States was estimated at 30–90%, depending on the area" (Smith 2011).

- "Enamel protects the nerves in your teeth. Enamel will never grow back once it is worn away, so these drinks, if not consumed correctly, will affect your tooth health forever" (Lee 2010).

- "More than 19,000 organ transplants are performed in the United States every year," reports *health-first.org*. However, approximately 106 people are added each day to the organ waiting list, and 18 people die every day waiting for an organ, as reported by the University of Pittsburgh Medical Center" (Ritter 2011).

Details, Details

With those first elements out of the way, the reader should know the big picture. Now the story moves into the details that illuminate what has been found or learned:

- "After 48 hours, my grandfather's chance of having a heart attack from smoking began to drop, all the nicotine in his body was out, and his sense of taste and smell will return to the normal level" (Vehige 2011).

- "The official rules of the International Tennis Federation state that in any professional tennis match with pressureless tennis balls, these tennis balls shall have an internal pressure no greater than 1 pounds per square inch (psi) or 7 kilopascals (kPa), compared to a regular ball which is 12 psi or 84 kPa" (Centeno 2010).

- "Creatures, such as eyeless shrimp and giant crabs, are also found around vents and get their food directly from black smokers. This is a process known as chemosynthesis" (Kelly 2011).

Comments and Balance

What do others think of this science, technology, or development? It is important to keep the story in balance:

- "If hippotherapy is so successful then why is it not more commonly used? According to Aetna Insurance Company, hippotherapy is still in

the investigational phase. They say there is not enough valid scientific data to support hippotherapy as a treatment strategy" (Goebel 2010).

- "However, there are disadvantages to landfill gas systems. Glenn O'Bryan, an engineer from Fred Weber, explains that landfill gas is an inferior fuel when compared to natural gas because it is only 50% combustible. Natural gas, on the other hand, is 100% combustible" (Bequette 2011).

- "'I agree that the school is currently inefficient and not very eco-friendly,' says School Principal Dr. Chris Greiner. 'But the new building that is currently being built [scheduled to be completed 2011] is much more efficient as far as lighting, and HVAC [heating, ventilation and air conditioning] with innovative features'" (Mian 2010).

The Triangle's Apex (or Bottom)

The story ends with sort of a whimper and no conclusion. It is typically a few sentences about next steps or related issues that are not critical to understanding the overall story.

- "Music therapy provides benefits to patients that are well tolerated, inexpensive, easy to manage, and free of side effects. These benefits are causing many to see music as an important tool to help the body in staying or becoming healthy, say experts" (Logan 2010).

- "According to *qmed.com*, a website supporting the medical device industry, the development of artificial organs, such as the artificial heart, kidney, liver, pancreas, and lung, is another revolution being investigated. This, too, is new science, and in most patients, artificial organs did not function as well as natural replacements" (Ritter 2011).

Signposts Along the Way

Interspersed throughout the story are elements that distinguish a good journalistic story. They will show up again as the basis of SciJourn standards (Chapter 4) and our definition of science literacy.

Attribution

The journalist's equivalent of references is the *attribution*, e.g., "according to.., says…, reports…." We find that many first-time writers have trouble figuring out first what to attribute and how. An FAQ on attribution can be found on *www.Teach4SciJourn.org*.

We typically say that anything that is not well known or well established should be attributed. Thus, the distance from the Earth to the Sun is not attributed, but the most recent count of sunspots is. The fact that increased risk of breast cancer can be inherited requires no attribution, but the most recent data on numbers of Americans diagnosed with the disease should be cited.

Numbers and Specificity

A lab report that states "We heated the solution for awhile until it changed color," should bring out a teacher's red pen, especially when compared to "We heated the water solution to 80°C for 2 hours until the color changed from light pink to blue." Similarly, the too often used "Recent findings show that the new therapy prevents skin cancer" sets editors' teeth on edge. The better sentence is "Washington University Medical School dermatologists reported in the September 15, 2011 issue of *New England Journal of Medicine* that the new treatment reduced the risk of certain types of skin cancer by 80%."

Experience suggests that getting students to be that precise is a challenge. It typically comes about, with prodding, in the second or third revision. However, the practice of capturing detail is an important lifelong skill and an important element of science literacy.

Moreover, every story has a number—what something costs, how many affected, units sold. We even urge students to do surveys and analyze their data as part of SciJourn (see Chapter 8 on surveys). And sometimes those numbers will need comparisons and analogies to make them meaningful and to put them into context.

Finally, getting into details is aided in the SciJourn approach because of the emphasis on narrow, local topics (Chapter 6 on finding an angle). It is difficult to talk meaningfully of huge, unwieldy issues such as cancer or global warming. Scientists and your writers need to focus on a small piece of these big topics to really say anything meaningful.

Reporting, not Editorializing

We often receive science stories in which the student authors offer their opinions or present one-sided arguments. Our response is that in a news article your job is to tell us what experts say and not what you think. Editorials are the newspaper's equivalent of the persuasive essay whereas the news article is more akin to expository writing. But remember, a good news article also tells a coherent story.

Hand-in-glove with telling a story goes reporting what experts, including opposing sides, think. A story that began in a student's mind as an argument in favor of natural grass versus artificial turf for sports fields evolved during the editing process into an examination of the relative benefits and problems of the two choices.

In an adult world, we hope that issues are nuanced. The question of what is safe, attainable, affordable, effective, and reasonable can lead credible experts to opposing or qualified answers. When journalists capture that, it is reporting. We were pleased when a story on alternative medical practices acknowledged their value, while warning that some don't work or may be dangerous.

Up-to-Date

Another phenomenon is that what is new to students can be old hat to scientists. A story that focused wholly on the dangers of volatile organic compounds (VOCs) in paints was deemed a dead end because VOC-free paints were already on the market. Those new paints are the real story and the VOC paints became the background information.

This can be tough for students used to operating in a textbook world. Science sometimes moves way too fast for even a "recent" report to contain the newest findings. An article on newly discovered arsenic-based microorganisms was already out of date by the first edit because other microbiologists were vigorously attacking the data.

5 Ws and Sometimes How

What every student seems to know about journalism are the five Ws—*who, what, where, when* and *why*. Some folks add *how* to that list.

Ironically, the five Ws are not something that journalists or editors mentally track. It sort of falls out of storytelling itself and emerges with the need for specificity. Yes, those Ws are most likely in every good news story—often multiple times. Good writing is not, as so many believe, a matter of getting the 5 Ws into the first sentence.

Early on in the project we did emphasize the 5 Ws, which led teachers to design exercises in which students picked out the *who, what, when, where,* and *why*—in effect turning writing into another set of fill-in-the-blanks. We suggest instead that authors and editors step back a level and ask themselves: "Is this a convincing story? Do I have the details I need?" The Ws come naturally out of those queries.

Breaking the Rules

By emphasizing storytelling, we see paths along which "breaking the rules" improves the story. Typically, we strip the "I" out of stories, but an article on migraines by someone suffering from the malady began with an "I" story of an emergency room visit. Similarly, we loved our Teen EMT series—real "I" stories of a high school senior who reported cases she encountered while working in an ambulance.

Although we rail against editorializing in stories, editorials, with care, can work under SciJourn rules. One chemistry teacher asked students to write a letter to the U.S. President regarding global warming action. His goal was to see if they could support their arguments with multiple, credible sources and accurate information.

These days, science journalists blog and podcast, comment and analyze legislation. Much of that would have been unthinkable just a few years ago.

Writing

We have deliberately downplayed the mechanics of writing in this book, even though journalists learn to write succinctly and clearly. By comparison, students tend to ramble on, oftentimes needlessly narrating their story (I then asked 10 students what they thought about....) or providing irrelevant details (In the late 1800s scientists first began searching for...).

We typically take care of these problems late in the editing process. Our first goal is to get the content (as defined in our standards described in Chapter 4) on the page, then model good writing or compare good writing to the rambling sort. What is good enough in terms of writing mechanics is the teacher's decision. What is good enough to be published is an editor's decision. SciJourn exists to push the boundaries on content.

References

Bequette, C. 2011. Converting trash to energy. *www.SciJourner.org*

Blum, D., M. Knudson, and R. M. Henig. 2006. *A field guide for science writers.* New York: Oxford Press.

Busch, K. 2, 2010. *Inside the iPad Revolution. www.SciJourner.org*

Centeno, I. October 10, 2010. *Why Airtight? www.SciJourner.org*

Emrick, D. July 20, 2010. *The Secret Number One. www.SciJourner.org*

Goebel, H. September 15, 2010. *Horses and Healing. www.SciJourner.org*

Haas, A. March 28, 2011. *Do Green Bags Really Work? www.SciJourner.org*

Kelly, K. June 10, 2011. *Are Black Smokers Architects of Life? www.SciJourner.org*

Kovach, B, and T. Rosenstiel. 2007. *Elements of Journalism.* New York City: Three Rivers Press.

Lee, M. December 8, 2010. *Don't Brush Your Teeth. www.SciJourner.org*

Logan, A. June 15, 2010. *The Mind-Music Connection. www.SciJourner.org*

Macheca, M. March 15, 2011. *The Sandman in the Classroom. www.SciJourner.org*

Mian, M. December 1, 2010. *How Green is Your School? www.SciJourner.org*

Ritter, R. January 25, 2011. *Recycle Yourself. www.SciJourner.org*

Smith, D. January 10, 2011. *Colony Collapse Disorder Not as Bad in Missouri. www. SciJourner.org*

Vehige, C. January 24, 2011. *Quitting "Cold Turkey." www.SciJourner.org*

CHAPTER 3

CAN I DO THIS? FREQUENTLY ASKED QUESTIONS

Teachers and administrators have expressed certain practical concerns, the answers to which help them decide if SciJourn is possible given the realities of their own school. In these Frequently Asked Questions (FAQs) we seek to address some of these worries, based on practitioners' experiences.

How Do You Find Time for SciJourn Given the Already Overloaded Curriculum?

No matter how carefully this book has been organized or how convincingly we argue for science journalism in the classroom, we understand that teachers operate in the real world, with real time constraints. What we can say, however, is that many teachers who struggle with these same issues—a prescribed curriculum; tests, tests, and more tests; a discomfort with teaching writing; a discomfort with science—have decided that it *is* worthwhile to spend time with science journalism. For many of them, the chance to relate what they know and care about to real-world concerns has become essential to their identity as teachers. And so they have made it a priority to find the time.

There seem to be several basic ways that teachers have folded SciJourn into their curriculum:

- *The mini-lesson.* Short lessons, something like 5–10 minutes a day, spread over a long period of time.

- *Dedicated, full-blown lessons spaced throughout the year* (e.g., SciJourn Fridays). Teachers spend time with lessons that result in a science news article (and sometimes even more than one!).

- *The "blitz."* Over the course of the year teachers may try two, three, or even four SciJourn Blitzes. This means that for one or two weeks at a time they work intensively with students to produce an article. Often it takes students more than one blitz to produce something usable, so students return to a topic they were working on as homework or during the next blitz period.

- *The collaborative approach.* The science and English or journalism teachers work together to build a science journalism program. The research may take place during science time and the writing during the English period. Alternatively, science students may write in one class and a different group of students peer edit in English class, or English students write the article and the science students fact-check it.

  In another instance, science teacher Linda Gaither submitted their first drafts to Becky Bubenik's journalism students who used the SciJourn standards to evaluate the submissions. Becky had the fledgling journalists begin by projecting a sample article on the board and the group brainstormed a list of issues to comment on. A few days later the newspaper students used the same list to respond to more articles from Linda's classes, but this time they worked in pairs. The overall response from the science students was positive and they used the peer edits to revise their stories.

- *Kill two birds with one stone.* In one parochial school, students were required to address a social justice theme, so the teacher embedded that requirement by assigning students the task of writing a news article that involved science and social justice. Helping students move from the editorial stance to a news article proved to be both challenging and productive. In addition to teaching students about the subject content, it also became a platform for exploring issues of genre and audience. In another school, students were assigned articles that required that they find contemporary, relevant information about an element on the periodic table. (See *www.Teach4SciJourn.org* for detailed accounts of these efforts.)

"Science teachers are really hungry to find a way to help their students connect content to their own lives and interests. Their curriculum often keeps students at arms' distance."—SciJourn teacher

- *A local science event or speaker.* An easy way to gather a story, talk to experts, and get reactions is to cover a science event (see Chapter 6: "What's Your Angle?" and Chapter 8: Original Reporting: Interviews and Surveys). The event can be taking place at the local science center or college, or during a school field trip. Town meetings on science-related issues are also an interesting territory to explore. The key is advance preparation in order to ask the questions needed to put together a story.

Teachers have also used invited speakers—for instance, an expert on hybrid cars or waste treatment—to launch a science journalism unit. In advance of the speaker's visit, the students do some research, listen and take notes while the guest is talking, and then look for an angle they want to use to follow up on a topic the speaker made them think about.

- *The no science fair option.* Teachers have given students the option of writing an article or completing a science fair project. The two activities seemed to take about the same amount of time.

- *The after-school option.* Several of our SciJourn teachers have opted to use the strategies and methods we suggest in after-school clubs or as an elective. Others are looking into the possibility of a summer school class. For several years we have offered SciJourn as one choice for local teens participating in the Youth Exploring Science Program at the Saint Louis Science Center.

Year-to-year, or semester-to-semester, teachers often change their approach and timing. Since our goal is not to produce a curriculum per se, teacher changes make perfect sense. Circumstances change, opportunities for collaboration wax and wane, and teachers find that the project works better in some classes than in others. What we have learned, though, is that even something as quick as Sci-Journ read-alouds (Chapter 5: Setting the Stage by Modeling) can make a huge

difference in terms of students' ability to think critically. It also appears that the more frequently SciJourn lessons are used, the more scientifically literate students seem to become, and the better their teachers become at responding to student work. Research to explore this correlation is currently ongoing.

Can SciJourn Fit the Curriculum?

When we met with the first group of SciJourn teacher participants, most identified their goal as "finding a way to support and add depth to the curriculum." To that end they sought to coordinate student topics with what was being taught in their biology, chemistry, physics, or Earth and space science classes, making fairly specific assignments. Assigning topics that match issues addressed in the curriculum seemed to show much promise; for instance, when studying the elements, a teacher might guide students to write about the authentic and current concern about rare earths. "Could students research issues related to Chinese and African use and export of specific chemical elements? Or perhaps issues related to helium might be easier for students to grasp," the teacher muses. Ties like these may anchor and make the study of the old tried-and-true topics come to life for students.

> "I find SciJourn rewarding because my students find it rewarding."
> —SciJourn teacher

One place where teachers find an easy fit between the curriculum and SciJourn practices is in what we call read-alouds/think-alouds that invite teachers to model their own understanding of text so that students get a holistic view of what happens when a scientifically literate adult reads. In an Earth science class, for example, the teacher might follow stories with her students about an event, such as the Russians drilling through two miles of Antarctic ice to sample a trapped, liquid water lake. In a physics class, students could follow stories about the latest planets found around distant stars.

Other teachers have tended to focus more on science processes. Two teachers, worried about the end-of-the-year high-stakes testing, asked students to read newspaper articles with graphs and identify the scientist's hypothesis, dependent and independent variable, what aspects of the experiment were kept constant, etc. And since human beings can do more than one thing at a time, the students also learned something about science content and newspaper form along with the science skills and processes the teacher had highlighted.

The teachers who had the easiest time imagining the addition of a science article to the curriculum were those who taught environmental science, forensics, or journalism, since these classes seemed to be more applied and the curriculum less jam-packed. What has happened over time, however, is that participating schools and teachers have recognized the value of science journalism as a way of showing the relevance of science to the lives of informed citizens and offered students more freedom in topic choice.

Regardless of approach, attending to "what's new," whether in chemistry, Earth science, or history, adds a dimension that can't be found in a textbook.

Each department in a high school, trying desperately to cover as much material as possible, has found its own way of "scaffolding" learning. However, in the process, have we actually removed thinking and problem solving from the curriculum? Teachers have sought a way to engage students, to help them see the personal and civic questions they struggle with informed by the rigorous sorting through of information that SciJourn calls for.

What Can I Do to Keep Students Motivated for Such a Long-Term Project?

Given the difficulties inherent in learning to read and understand science as well as students' oft-repeated aversion to writing, SciJourn teachers realize that students need to be motivated. Here is the key: interest is a huge motivator, more effective than monetary rewards, good grades or almost anything else you can imagine. But how many students are able to find topics and specific examples that motivate them in their basic textbooks? Do students currently find their reading in science important? Engaging? Pleasurable to pursue? Are they at any point free to choose science topics of interest? Do our science classes help young people with the basic work of adolescence (i.e. developing their identities)? How might we bring science journalism to them in small (and gradually increasing) bits?

Research-based pedagogical strategies that seek to support student engagement and learning (Bransford et al. 2000; Micheals et al. 2008) have provided a useful frame for thinking about motivation. Experts say we need:

- *Intellectually and socially authentic activity*. Students work as journalists, rather than listen to a teacher lecture. By becoming journalists, students are responsible for identifying questions, finding sources, evaluating credibility and writing a final product. The teacher provides scaffolding as necessary, but the student is ultimately the worker.

- *Valuing community and distributed expertise*. In almost all of the lessons in this book, students must work through problems to discover and assess potential solutions. Problems range from finding a suitable topic to identifying "experts" to handling contradictory evidence.

- *Cooperative learning*. Although this book generally focuses on articles with single authors, lessons encourage students to work together in other ways. Nearly all the lessons call for a group problem-oriented discourse. We have also included lessons on peer editing and creating classroom resources.

- *Open-ended, student-centered, classroom discussions*. The discussions that arise from these lessons are not the kinds of "discussions" in which the teacher already knows the answer. In many lessons, students bring in articles or information that the teacher has never seen before. In other lessons, problems or scenarios that have no single correct answer are proposed. In all cases, student voices are central to the learning.

- *Building on/from student interests.* While teachers may choose to use this approach in different ways, once students begin working on their own articles, their questions will drive instruction. Teachers may decide to limit topic choice, but students will have to develop their own questions within these topics if they are to truly act as journalists.

- *Real-world problem solving.* Because science journalism is a real field with many real-world examples to draw upon, students are working on authentic tasks. In addition, we have provided suggestions for publication so that student writing may gain the added authenticity that comes from having a real audience.

This brings us back to SciJourn. If one looks at the topics that students have chosen to write about—hippotherapy, cochlear implants, Potter's syndrome, the timing of penalty kicks—it is clear that neither textbooks nor the curriculum would offer them much information about these subjects that has deep personal meaning and which motivates students to dig further, understand more, and put ideas together. In an era when reading levels are used to determine much of what is offered to students as text, it is refreshing to see these students have an opportunity to struggle with personally meaningful information, motivated by a need to know more. We may be reminded of our own efforts to make our way through medical texts, surely above our reading level, to determine if a symptom is real or imagined.

We have also found that students are motivated by the success of their peers. The first story we received from a struggling high school was from a less-than-high-achieving young man in an advanced chemistry class. What originally got him going was a sense of pride and interest, but by the time the editor received his first story it had been through nine drafts… and it still wasn't great. Our editor continued to work with him, and finally, the first story from the struggling school was published. Of course, other students learned about this student's success and other teachers learned of the success of his chemistry teacher. "If they can do it, I can do it, too." This school probably has more published stories now than any of our other sites. They, and we, are proud of their achievement. Success of this kind also adds to confidence.

Do I Have the Needed Technology?

SciJourn depends largely on students' access to the internet. In many schools we visit there are computer labs, but access to those facilities is limited. The preferable situation is to have in-class access to the web—something that we hope will be the norm in the not-too-distant future.

There are a few teachers who have worked around technological challenges by sending students to the local library or university, assuming they can do computer-based assignments at home, or by purchasing a wireless modem. The modem at least allows a teacher to do read-alouds or show articles on an overhead or white board.

The far more common problem, however, has to do with access to information. Schools have blocked not only sites, but all uses of certain words. Try doing a story on breast or anal cancer when your key terms are censored! Similarly, in many schools with excellent computer access, students are not allowed to use e-mail, which makes setting up an interview or interviewing an expert online nigh impossible. Our best advice is to anticipate these problems ahead of time so that you don't find yourself in the middle of a project, arguing with the principal or tech specialist as the students wait to proceed with their articles.

Is It OK to Work Outside My Area of Expertise?

Whether a function of personality or a philosophy of teaching, some teachers are very uncomfortable not knowing more than their students; they believe it is their job or responsibility to be able to offer those in their charge advice and information. In SciJourn, the goal is for teachers to understand more about the process of finding and using credible information, but not necessarily know more about the content.

We have found that there is a difference, however, between science teachers and journalism teachers who have engaged in this project: the science teachers are more able to spot factual errors, but this happens largely with respect to background information or context. We offer a number of hints in Chapter 9: Channeling Your Inner Science Teacher: Considering Context and Accuract, which are designed to help teachers help their students. Our best advice is to remember that students get better at finding and assessing information with experience. The same is true for teachers.

Students in Our School Have Been Taught to Write Using the Five-Paragraph Essay. How Does That Relate to SciJourn Writing?

The five-paragraph essay, for those unfamiliar with the term, is a form often used by school systems to train students to write in such a way that evaluators will give full credit for essay answers. Typically, the student is taught to make an assertion (e.g., dogs can be an important support to humans). Then the student offers three supporting details (e.g., dogs have been used to help the blind, to treat post-traumatic stress disorder, and to help abused children). Then, the student summarizes the points in a concluding paragraph (e.g., dogs really are man's best friend).

First, it is important to notice in this and similar examples that the topic is much broader than anything found in a *SciJourner* article. But there is another

"[SciJourn] was just worth it all because for every kid that got published there was another kid who was inspired to do a little bit more."
—SciJourn teacher

problem as well; the five-paragraph essay is typically about structure and not much about logic. Students tend to look for three paragraphs to plug into the formula without thinking about connections between and among the examples. In this case of the dog essay, there is only a broad connection to the opening assertion and virtually no connection between the examples, i.e., the three uses of dogs. A news article, by contrast, demands logical thinking. To talk about support dogs with abused children, one would have to understand how support dogs are being used. How are they trained? How many abused children are being helped? How do we know it helps? Where has this been done? What other treatments are available?

Clearly the structure of an essay—beginning with a thesis, supporting that thesis, and ending with a conclusion—is much like the kind of writing expected in literature and history courses in college. However, these academic essays rely on logic absent from the five-paragraph essay in its diminutive form. Still, students *are* learning the form; students often submit news articles—even after weeks of training—that look just like the five-paragraph essay. If this happens in your classes try a lesson that helps students compare the two forms (see our writing and revision chapters, 10 and 11, or *Teach4SciJourn.org* for exercises that may need to be repeated and referred to again and again).

Who Gets Published? Is That a Good Thing?

Teen writers seem to inspire the next generation of teen journalists and that is why we publish the online news magazine, *SciJourner.org*. The other reason is that the website offers examples for teachers and students, models that can be useful in various ways. These articles have "worked."

In our implementation of the project, those students who were published were those most engaged, not those who are necessarily the best writers or traditionally the best students. Some teachers have worked with students through multiple drafts, accepted "creaky" writing when the content was good and guided authors to credible sources, all in an effort to get a publishable story.

What is good enough to be published is a difficult bar to place and depends on the editor. However, we strongly reject the idea of publishing stories to make students feel good or because they met some arbitrary goal such as one revision. Authenticity means that the story should, in the eyes of the editor, meet high standards.

Please note that publishing an article in the science section of the school newspaper or even in the local paper (yes, some of our communities have allocated space for teen journalists if their articles are good enough!) is just as rewarding. And sometimes just the act of gathering information to address a science-based curiosity (Can you really erase memories? What happens to

donated organs?) is often enough to spur students to learn and think more critically and scientifically. Having an opportunity to share their insights, as well as information that they have decided *not* to use, with classmates and get "credit" for this sharing is an added benefit.

Students who aren't published still benefit from being exposed to the Sci-Journ standards. We have seen over and over classrooms adopt language on credibility, multiple sources, and context, for example, after going through the project. Some students who don't complete a story initially may even strike out on their own to create a new, better story later in hopes of being published. The key, again, is personal engagement with a topic.

Can My Students Handle This?

Two experienced biology teachers with whom we work came up with an interesting alternative to the science article. They have their students construct PowerPoint presentations instead (see Chapter 6: "What's My Angle?"). Some of the motivated students have even turned their PowerPoints into submitted pieces. This seems to serve well as a kind of stepping stone. Other teachers have differentiated by having some students do photos with captions or voiceover slide shows. More suggestions like this can be found in Chapter 12: Beyond Words.

What About When the Topic Makes Me Uncomfortable or Ethical Issues Might Arise?

Some science stories, such as drug use, sexually transmitted diseases, or teen pregnancy, can take you and the author into some tough places, especially if the student is collecting school data or conducting their own interviews. Many schools have rules that would make these topics *verboten*, even if the interviews and surveys are anonymous. The worry is that some names might be leaked. It is important to check with your administration and lay down rules at the beginning on what topics are off-limits. (If you follow our approach, you will be tasked as editor with approving topics before the writing begins.)

Other teachers have simply said no articles on evolution, abortion, stem cells, etc., arguing that nothing anyone will write will sway opinions on polarizing topics, and they simply don't want to fight those battles. In our experience, these are not the topics teens want to write about anyway, since they usually lack a personal connection.

However, we have run into several student-written health stories that raise ethical issues. For example, stories in which the student reveals personal or family medical history, admits to not receiving treatment for a medical condition, or suggests that the school or another institution is turning a blind eye to a dangerous practice. The last item has often taken the form of a sports story in which players admit to multiple concussions or a cover-up of a contagious medical problem.

Journalists would consider all of these good stories but might protect sources by not naming them. Moreover, an extra burden falls on sources under

18 years old (see Chapter 8: Original Reporting: Interviews and Surveys). We always check that adult family members are OK with being named—many are—and protect those who do not wish to be identified. The other issues, such as inadequate health coverage and school problems, require a conversation with the school administration. What we don't do is leave a student exposed to possible repercussions from a story, just as a newspaper stands behind its reporters.

I Would Like to Keep Track of What My Students Have Learned. Any Ideas?

To better understand the meaning of their SciJourn work from a student perspective we offer two concrete suggestions; one involves individual learning logs and the other is more akin to a KWL chart created by the class. For the learning log to work well, you must allocate time for students to record their thoughts, questions, setbacks, and successes as they go along. Alternately, the class—under the teacher's direction—could keep a large three-column chart in the classroom. In the left column, students list what they already know or do in relationship to the different topics covered in this book. For example: What do they know at the beginning of the unit about the topic at hand (e.g., How do they currently search for information? or What do they know about source credibility?). In the middle column, students list their questions or problems, attending especially to the way they currently work. This list can be expanded as students encounter new problems. In the right column, students write, perhaps as part of a class summary at the end of the unit, the pertinent things they have learned. The result would offer tangible evidence of what students have gained from their SciJourn work.

Can SciJourn Work as a Group Project?

So many teachers have wanted to see science writing as a cooperative learning activity. This is challenging because at some point, someone needs to be the lead writer. We have seen successful teams in which members are given assignments, such as an interview, collecting targeted data, or taking photographs, but the job of putting it all together and making it work, has always fallen to one person.

Do I Have to Follow This Book in Order?

You can pick your path through the book and use chapters to reference issues as they arise. The book's order follows how a typical SciJourn project unfolds and chapters highlight the important issues that often arise.

I'm Sold! What's the Most Important Thing to Do Before Getting Started?

You're not going to believe this, but by far the most important thing you can do is to write an article yourself. Good teachers don't have students do labs they haven't tried, right?

References

Bransford, J. D., A. L. Brown, and R. R. Cocking. 2000. *How people learn: Brain, mind, experience, and school.* Washington, DC: National Academies Press.

Micheals, S., A. W. Shouse, and H. A. Schweingruber. 2008. *Ready, set, science! Putting research to work in K–8 science classrooms.* Washington, DC: National Academies Press.

CHAPTER 4

SCIENCE JOURNALISM STANDARDS

L et's face it: schools are under attack. As the narrative goes, we in education are not doing enough and we're doing it badly. In an attempt to address the oft-cited curricular failures evident in the world of education—as well as issues related to national identity and demographic shifts—virtually every professional and many political organizations have called for the establishment of "standards." At this point in history, state and national standards, to a large extent, both determine the curriculum and set benchmarks for acceptable student and teacher performance.

Standards can be used in two basic ways: as aspirational goals to establish hopes for our students or as gatekeepers to measure what is acceptable performance. We suggest that classroom teachers view our standards as primarily aspirational—a target for students to try to reach, an articulation of what we count as important when a student tries to tell a science story in a scientifically literate way. At the most basic level, the SciJourn standards summarize and reflect the project's values, the values that you have read about in earlier chapters, which are further elaborated and made concrete in the remaining chapters of this book. Or you could think of them as the "elevator speech," the quick and dirty version of what this project is about.

Because the content students write about is so varied, the subject matter standards used in your school probably do not guide you to become a better SciJourn teacher. Said differently, although the information students gain by writing an article can typically be tied to one content standard or another in biology, chemistry, or physics, for instance, it is frankly impossible to list all of the content learning one might expect to occur through writing science news. What we can do, however, is to be clear about the underlying principles that we use to focus, organize, describe, and assess activity.

Our goal of encouraging students to act as science journalists is mostly to help them become more scientifically literate. Remember, our definition of science literacy has to do with teaching students skills that will have utility 15 years after high school, when science information will almost surely change. With this in mind, we define scientifically literate individuals as those who are able to:

1. Identify personal and civic concerns that benefit from scientific and technological understanding.

2. Effectively search for and recognize relevant, credible information.

3. Digest, present, and properly attribute information from multiple, credible sources.

4. Contextualize technologies and discoveries, differentiating between the widely accepted and the emergent; attending to the nature, limits, and risks of a discovery; and integrating information into broader policy and lifestyle choices.

5. Fact-check both big ideas and scientific details.

Our SciJourn writing standards mirror and build upon these science literacy goals.

In this chapter, we briefly outline the process used to create the SciJourn standards, and then describe each in some detail, furnishing the reader with real student examples. As you seek to "sell" the project to your schools, parents, and community members, it might be useful to connect SciJourn goals and standards to those created by national boards. It may, in fact, be useful for you to consider developing your own list of connections between SciJourn goals and the standards your school relies on. Figure 4.1 provides an example, highlighting the connections between SciJourn standards and those identified by the NRC.

FIGURE 4.1

The National Science Education Standards* vs. SciJourn Standards

(a) Fundamental abilities necessary to do scientific inquiry	Adapted for science journalism
▶ Use data to construct a reasonable explanation.	▶ Science journalists use both firsthand (investigative reporting) and secondhand data to construct reasonable explanations.
▶ Communicate investigations and explanations.	▶ Science journalists communicate investigations and explanations.
▶ Develop descriptions, explanations, predictions and models using evidence.	▶ Science journalists describe, explain and predict the importance of scientific findings. They create analogies, which sometimes function as models.
▶ Think critically and logically to make the relationships between evidence and explanations.	▶ Science journalists think critically and logically to make the relationships between evidence and explanations.
▶ Communicate scientific procedures and explanations.	▶ Science journalists communicate scientific procedures and explanations.
▶ Identify questions and concepts that guide scientific investigations.	▶ Science journalists identify questions and concepts that guide scientific investigations.
▶ Recognize and analyze alternative explanations and models.	▶ Science journalists recognize and analyze alternative explanations and models.
▶ Communicate and defend a scientific argument.	▶ Science journalists communicate scientific arguments and the evidence used to defend them. *Continued*

* National Research Council (NRC) 2000. *Inquiry and the national science education standards: A guide for teaching and learning.*
Washington DC: National Academies Press.

Figure 4.1. The National Science Education Standards vs. SciJourn Standards *(continued)*

(b) Fundamental understandings about scientific inquiry	Adapted for science journalism
▶ Scientists develop explanations using observations (evidence) and what they already know about the world (scientific knowledge).	▶ Science journalists report explanations highlighting observations (evidence) and contextualize new information based on what is already known about the world (scientific knowledge).
▶ Scientists make the results of their investigations public; they describe the investigations in ways that enable others to repeat the investigations.	▶ Science journalists make the results of investigations public; they describe the investigations in ways that enable others to understand and track down others' investigations.
▶ Scientists review and ask questions about the results of other scientists' work.	▶ Science journalists review and ask questions about the results of scientists' work.
▶ Scientific explanations emphasize evidence, have logically consistent arguments, and use scientific principles, models, and theories.	▶ Explanations provided by science journalists are based on the assessment of evidence, the logical consistency of arguments and the use scientific principles, models and theories.
▶ Science advances through legitimate skepticism.	▶ Science journalists evidence legitimate skepticism as they present ideas and information.
▶ Scientists usually inquire about how physical, living or designed systems function.	▶ Science journalists report on how physical, living, or designed systems function.
▶ Scientific explanations must adhere to criteria such as: a proposed explanation must be logically consistent; it must abide by the rules of evidence; it must be open to questions on possible modifications; and it must be based on historical and current scientific knowledge.	▶ The explanations of science journalists (in a well-written article) adhere to criteria such as: a proposed explanation is logically consistent; it abides by the rules of evidence; it is open to questions on possible modifications; and it is based on historical and current (up-to-date) scientific knowledge.
▶ Results of scientific inquiry—new knowledge and methods—emerge from different types of investigations and public communication among scientists.	▶ Science journalists report on new knowledge and methods that emerge from different types of investigations and public communication among scientists.

How the SciJourn Standards Were Developed

Writing, whether it is in the science, history, or English classroom, is generally thought of as a literacy or literacy-related activity. The literacy community, like the science community, has created standards that seek to measure student and teacher achievement. Because the literacy community sees itself teaching the *skills* of writing and thinking, we first turned to well-regarded writing standards to see if they might serve the SciJourn project well. The 6+1 traits scoring rubrics (Northwest Regional Educational Laboratory n.d.) are the most commonly used assessment measures and are designed to help teachers assess student work by focusing on:

- *Ideas*, the main message;

- *Organization*, the internal structure of the piece;

- *Voice*, the personal tone and flavor of the author's message;

- *Word choice*, the vocabulary a writer chooses to convey meaning;

- *Sentence fluency*, the rhythm and flow of the language;

- *Conventions*, the mechanical correctness; and

- *Presentation*, how the writing actually looks on the page.

Given that SciJourn is primarily interested in content, we looked most closely at the part of 6+1 called ideas. The elaborated ideas criteria include:

- The main idea is clear, supported, and enriched by relevant anecdotes and details.

- Topic is narrow, manageable, and focused.

- Support is strong, credible; uses resources that are relevant and accurate.

- Details are relevant, telling; quality details go beyond obvious and are not predictable.

- Author writes from own knowledge/experience; ideas are fresh, original, and uniquely the author's.

- Reader's questions are all answered.

- Author helps reader make many connections by sharing significant insights into life.

Although we can fit many of our SciJourn concerns into one or another of these categories, the 6+1 rubric simply does not prioritize what is needed in order to write an exceptionally good science news article. For example, in a 500-word science story the writer must make choices about what to include and what to eliminate. A good anecdote that is engaging but does not help explain the science may, in fact, be distracting and a waste of words. On the other hand, not including details or information on why a particular finding is important is sure to kill the piece.

It was clear to us that the 6+1 writing traits rubric was not going to serve our purposes. To develop a set of standards that could measure the kinds of science literacy skills we believe are important, we next turned to relevant stakeholders including science journalists and editors, literacy and science experts, teachers, and their students. Bringing an authentic, relevant, out-of-school practice into the classroom involves lots of people who evidence different kinds of knowing; negotiating what is most important to each is no easy task.

We began by listening to science journalists as they talked about their own writing. How did they go about doing their work? We learned that they valued their ability to locate information, to ask good questions, and explain what they found in ways that helped answer questions they were able to anticipate. Equally productive were our discussions with science editors. They generally saw themselves as a proxy for their readers. Using that lens they were able to ask writers for clarification, push them to gather more information, and encourage them to explain things more clearly. Our SciJourn goal was not to hold fledgling authors to these same standards, but rather to understand the task of a professional so that we could scaffold the experience of writing for a student audience.

We also met with folks who were undoubtedly scientifically literate and those who were not. What did a physicist do when reading outside his field? What makes something clear for one reader and almost impossible to another? Is it just vocabulary? Is there something in the way the information is explained? What makes a concept easy to understand?

We then took what we had learned from the professionals and brought our knowledge into classrooms … and made a lot of mistakes! For instance, we thought that it might be a good idea to begin by getting students to pitch their ideas, just the way journalists do in a newsroom (more about that in Chapter 6: "What's My Angle?"). We spent literally hours listening to and honing pitches. Not that it was a bad exercise, but it simply took too long, especially since we wanted students to have time to write and revise. Instead of beginning by teaching about who, what, where, when, and why, we soon realized that there are more fundamental issues, like learning how to search for and evaluate websites. Searching, it turned out, was not only a critical step in writing a story but also a critically important part of becoming more scientifically literate. In retrospect this seems so obvious, but it took us time to realize that not spending enough time searching, or pushing prematurely to the written product, really does a terrible disservice to the science literacy aspect of science journalism.

Figure 4.2 provides a one-page version of the SciJourn article writing standards that mirror the science literacy standards cited on page 36. We have also developed what we call a "kid friendly" version of the standards that students

might find less threatening or more natural (see Figure 4.3). What follows is the rationale for each of these standards as well as examples you should feel free to use in your teaching.

Figure 4.2

SciJourn Article Writing Standards

A science news article is a tangible display of scientific literacy. A good *SciJourner* student article

I. Has most or all of these elements: local, narrow, focused, timely, and presents a unique angle.

   a. Findings are meaningfully applied to personal or civic issues.

   b. Readers' likely questions are anticipated and addressed.

II. Uses information from relevant, credible sources including the Internet and interviews.

   a. Successful authors use internet search terms and search engines effectively.

   b. Successful authors privilege data from credible government and nonprofit sites and can justify the use of "other" sites.

   c. Successful authors locate and query experts and relevant stakeholders.

III. Is based on multiple, credible, attributed sources.

   a. Sources are relevant and reliable.

   b. Stakeholders with varying expertise and experiences are consulted.

   c. Sources are identified and basis of expertise is explained.

   d. All assertions, numbers, details and opinions are attributed.

IV. Contextualizes information.

   a. The article tells why the information presented is important both from a scientific and societal viewpoint.

   b. The article indicates which ideas are widely accepted and which are preliminary.

V. Is factually accurate and forefronts important information.

   a. Science connection is evident.

   b. Difficult concepts are explained.

   c. Precise language is employed.

   d. Quantitative measures are given in correct and comparable units.

   e. Information is up-to-date.

   f. Captions and graphics are checked for accuracy.

The SciJourn Standards

For our SciJourn teachers, a science news article is not just a writing assignment: A science news article is a *tangible display* of science literacy. Each of the qualities we associate with a scientifically literate individual can be made manifest in the article.

<u>Standard I</u>: A good *SciJourner* student article has most or all of these elements: local, narrow, focused, timely, and presents a unique angle.

Reporters have a duty to address the interests of their audience and that often means reporting on current issues that are of local relevance on topics that the audience cannot find elsewhere. Even nationally read newspapers, such as the *New York Times* or *USA Today*, include local news. In addition, news is by definition narrow and focused; books and feature-length (3,000- to 5,000-word) magazine articles have the luxury of taking a bigger view of a topic.

Perhaps more important, there is a practical side for your students in this standard. Broad topics lead to vague and poorly researched articles at best, or a series of clichéd ideas at worst. Imagine 500 words on "air pollution is bad" versus "city smog levels soar this summer." Clearly, the latter, local, narrow topic requires the author to go into some science depth in order to tell the story. We also expect that the story stays focused on smog and the city, and doesn't wander into global climate change or an editorial about the problems with cars.

Moreover, by encouraging the author to find a unique angle to the story, we push the student to do something original. What could our student journalist do to make her story unique? Interview neighbors on how the smog levels affect them, recognizing that older folks are more susceptible to the bad air, perhaps? An easier and more teen-relevant interview might be to discuss the topic with the high school track coach or a teen with severe asthma. You'll find lots more ideas on finding appropriate topics in Chapter 6.

Here are some *SciJourner* examples that bring interesting aspects of this standard to light.

- A 17-year-old writes about how her training as an emergency medical technician. She describes her work in an ambulance and how she is informed "of the patient's current condition and how to treat that condition to the best of my ability" (Turner 2009). This story and others like it show how science and technology affect each of us personally.

- "Have you ever wondered why every time you open a brand new can of tennis balls you hear a 'pop!' sound?" (Centeno 2010). This story, though not particularly local, is narrow and unusual and anticipates the readers' likely questions. As a reporter this student needed to ask (and understand) the difficult science questions.

- We have had several good stories on sports injuries, but this teen writer highlights a narrow medical issue of great personal concern

and finds a timely, local angle to the problem: "On July 13, Gov. Jay Nixon signed the Interscholastic Youth Sports Brain Injury Prevention Act, which is better known as the Missouri Concussions Bill" (Hawkins 2011). The author could not do a good job on this story without appreciating what a concussion is medically and the associated health risks.

Standard II: A good *SciJourner* student article uses information from relevant, credible sources, including the internet and interviews.

Reporters are expected to research their subject before writing a story, collecting background information, identifying credible sources and exploring the issues and controversies surrounding a topic. Their job is then to build a story by weaving together appropriate information from various sources, including interviews of appropriate experts and stakeholders—individuals with a personal, financial, or political interest in the discovery or invention. Controversies are handled by quoting relevant experts, not by choosing one side over the other. However, false controversies are avoided by recognizing who is a credible expert and what is a credible source… and who and what is not.

For our student writers, this is a tough standard. As noted in Chapter 1, *credible* in school has by-and-large meant the textbook or the teacher's lecture. The internet is filled with sites that may look credible but are not. We all need to learn to identify those credible sites that keep their information up-to-date and maintain quality control over their material.

Reporters typically rely on government and nonprofit websites, but there are many useful sites that will fall into the "other" category. The story's topic, to a large extent, determines what is credible. In looking at information on hybrid cars, comparing car manufacturers' sites is an important thing to do.

Finding the right terms also makes a huge difference. For example, "astrobiology" as a search term returns more credible sites than "life on other planets" because it is a word used by scientists. Chapter 7: Finding and Keeping Track of Sources, goes into greater detail about the challenges of and strategies for attending to both of these challenges.

Successful authors also need to locate and query experts and relevant stakeholders. Chapter 8: Original Reporting: Interviews and Surveys provides ideas on how to find and interview local experts and stakeholders.

Here are some *SciJourner* articles that meet this standard:

- "According to the National Institutes of Health and Mayo Clinic, a migraine may be caused by imbalances in brain chemicals called neurotransmitters" (Ragland 2010). It is our experience that most teens have never heard of either NIH or Mayo Clinic, so this inclusion in the story reflects learned information and effective searching.

- "Fetuses afflicted with Potter's Syndrome lack one or both kidneys before birth, according to *potterssyndrome.org*, a website run by a mother who lost a child to the syndrome" (Pool 2010). The cited website in this article falls into the "other" category, so the author included language to help the reader judge the credibility of *potterssyndrome.org*.

- In a story about her sister's impending blindness, one teen wrote: "When she (Diana) reads, she must have the reading material not even a nose length away. Diana is only 9 years old and the doctors of Children's Hospital in St. Louis have given her approximately 5–7 years to have sight" (Harris 2011). This article pivots on a dramatic interview with a stakeholder, a young girl with congenital glaucoma. Stakeholders can also be environmentalists, politicians, business leaders or the neighbor next door, depending on the topic.

Standard III: A good *SciJourner* student article is based on multiple, credible, attributed sources.

The goal of this standard is to recognize that science is an ongoing discussion and that one needs various perspectives to help inform the research process as well as public understanding. A scientifically literate individual further understands that even credible sources reflect biases or leanings; this is yet another reason to favor multiple sources. We want students to understand that any single source of information, because it *is* a single source, is limited.

Students also need to understand what can and cannot be generalized from any given source. For example, quoting U.S. data from the Centers for Disease Control and Prevention, despite the fact that it is a reliable source, is not as good a source as the World Health Organization for a story on AIDS in Africa.

Sources are identified in news stories through attribution, which recognizes that information has a source (who/which may have a certain agenda), and provides a pathway for the reader to verify and expand on something in the story (just as science journal articles must provide sufficient information to replicate the experiments). The attribution also establishes a historical record of where an opinion or concept started. Less formal than a reference, an attribution typically includes an individual name or name of the organizations, website, newspapers, TV shows, reports, and press releases. In many cases the citing of the name is followed by that person's or organization's reason for being cited as an expert.

Attribution is particularly important today. How many of us have heard "talk radio" callers or hosts or information from a "high school friend" repeated as fact? We find it gratifying when we hear a student ask, "What's your source?" or "Who says?"

Familiarity with journalistic and academic conventions is key to the understanding and use of attribution. The details included in the attribution help the reader form an opinion on whether the information is trustworthy. In some cases, it may mean understanding who is behind the work of a researcher or

organization. Students who pay attention to details such as who supports or funds certain types of work are well on their way to becoming scientifically literate.

- In a *SciJourner* story on tattoos, this student simply laid out two differing opinions, not choosing one side over the other and inviting readers to make their own decisions: "If you go to a scratcher you run the risk of getting hepatitis. As far as getting HIV from tattooing, well it's sort of a myth, claims Oldershaw. [Oldershaw is identified earlier in the story as a tattoo artist with eight years of experience, which makes him an expert.] Centers for Disease Control does declare that the risk still exists in getting HIV from tattoos" (Stavri 2009).

- "'Human behavior has a big impact on disease spread. And virtual worlds offer an excellent platform for studying human behavior,' said Professor Nina Fefferman in a 2007 interview with the BBC. Fefferman is a researcher at Tufts University School of Medicine who specializes in studying viral diseases" (Whistler-Brown 2010). This snippet shows a citation within a citation. The quote is attributed to a professor with strong expertise, but the words were taken from another credible source, the BBC. The 2007 date helps determine if the comments are current.

Standard IV: A good *SciJourner* student article contextualizes information.

Context, which includes social, ethical, economic, and political factors as well as science background, puts the story in perspective. Context also helps the author and readers understand why they should care about the discovery or technology and why researchers are interested in the topic. In addition, context underscores the interconnectedness of science and society and the cumulative nature of scientific research. SciJourn students are asked to understand the nature, limits, and risks of a discovery, emerging concept, or technology. Questions such as "Does the new knowledge change how experts view the topic?" or "Does it confirm what is known and believed?" help SciJourn students attend to context-related issues.

- "According to the Missouri Department of Health and Senior Services, the St. Louis Region reported 5,294 cases of chlamydia in 2008. Around 70% of those cases involved women infected with chlamydia" (Emrick 2010). The "classic" background paragraph is a cornerstone of virtually all health stories—numbers afflicted, numbers that may die, which groups are affected, and regional differences. It immediately indicates whether this is a rare malady or one that strikes large or special populations. Notice also that the data is tagged with a date; we encourage teen authors to find the latest statistics. This author is tackling a potentially off-limits subject in some schools, the spread of the sexually transmitted disease chlamydia.

- "Have you ever picked up a bottle of water from a convenience store when you were feeling thirsty? If you have answered yes, then you are one of the Americans, who combined, drink approximately 8.5 billion gallons of bottled water every year, according to the International Bottled Water Association" (Distler 2011). Although the author's story is about whether bottled water is better than what comes from the tap, the context paragraph is necessary in order to indicate the popularity of a consumer product.

- "If hippotherapy is so successful then why is it not more commonly used? According to Aetna Insurance Company, hippotherapy is still in the investigational phase. They say there is not enough valid scientific data to support hippotherapy as a treatment strategy" (Goebel 2010). The author accomplished something sophisticated in this paragraph. After a long explanation of the benefits of hippotherapy—how horseback riding helps people with certain maladies—she alerts the reader that this is still not a well accepted therapy, at least from the perspective of insurance companies. Teens who often think that almost anyone with an advanced degree is trustworthy learn through SciJourn that science is fundamentally cautious and in this sense conservative. This standard specifically asks readers and writers to beware of the word "proven."

Standard V: A good *SciJourner* student article is factually accurate and forefronts important information.

This standard is really about getting students to (1) pay attention to details by checking facts for accuracy and attribution and (2) understand the science well enough to explain it to someone else. In this sense it is about knowing what in the article is important and making sure that the reader understands that as well.

The 5 Ws of journalism that virtually all our students come to the project knowing—who, what, where, when, and why—are really about the details. But like any helpful reminder, they will not guarantee a good story. What will help guarantee a good story is being able to state the big idea in a succinct manner. This will show the teacher—and the readers—that the author "got it." Looking to see if they understand is fairly easy since reporters state the gist of an article in one of the early paragraphs.

It surely helps if the writer understands the scientific inquiry methods and scientific processes on which she or he is reporting. The student author should provide sufficient information so that the reader understands the finding and how scientists arrived at it. This requires the student to understand and digest the technical elements of the research and to use precise language (but not just science jargon).

The reporter's job, and the author's challenge, is to explain scientific ideas simply without changing the science. Consider the problem of astronauts

"floating" in space vs. living in a low-gravity environment. Or discussing the disease diabetes and never differentiating between types 1 and 2. In the long run, the new discovery or technology may be incorrect or fail (e.g., cold fusion), but the initial reporting should describe the situation as accurately as possible.

Students also have trouble with another accuracy-related issue: Many don't understand how numbers work. The number of people diagnosed in a given year is not the same as the number who die from a disease. When you interview four people, giving results as a percentage makes little sense. Units of measure need to be comparable. Very large or very small numbers are typically made more meaningful with analogies or comparisons to help the reader comprehend the value.

Finally, accuracy means that the information is up-to-date. Reporters strive to break news, but for teen authors being up-to-date is sufficient. We often see teens cite numbers that may be 10 years old, and we push back with the question of whether there are newer numbers. The same can be said for research results. A six-year-old study is not a recent finding, especially in a dynamic area of research. No one, for instance, wants to promote a medical treatment that has been discredited. An interest in timelines encourages students to look at publication or announcement dates as a means to determine whether it is up-to-date.

- "Cardiac arrest was the coroner's announcement following the tragic death of music legend Michael Jackson on June 25. Many people believe that meant a heart attack, but going into cardiac arrest is something different" (Johnson 2009). The hook for the story is the death of Jackson, but the author's real goal is to clear up a misconception. The science is not new, but the connection to Jackson's death makes the story timely. Moreover, the writer has clearly stated the gist of the story—people get heart attacks and cardiac arrest mixed up and I'm going to explain the difference.

- "In June of 2010, the Environmental Working Group (EWG)—a team of scientists, engineers, policy experts, and lawyers that reviews the work of the government—published an analysis concluding that a form of Vitamin A, called retinyl palmitate, found in approximately 41% of sunscreens may accelerate the development of cancerous skin tumors and lesions when applied to the skin" (Jacobsmeyer 2011). This article discusses a recent controversy. A latter section says the following:

- "There are two types of UV radiation, UVA and UVB, that could cause damage to the skin and cause an increase to skin cancer. UVA wavelengths are long wavelengths and range between 320–400 nanometers (nm); UVB wavelengths are short wavelengths and measure between 290–320 nm. UVB rays are the chief cause of sunburn, whereas UVA rays penetrate the skin more deeply and can cause the skin to wrinkle and look leathery" (Jacobsmeyer 2011). Although this section could have been taken from a textbook, the story centers on the controversy over vitamin A in sunscreen and not on general facts. It makes this background section seem more relevant and engaging. This section also demonstrates how nearly every story has a number—a percentage, cost, patients tested and so on. Citing those numbers is an important element of scientific practice and science writing.

SciJourn Ethics

Standards typically focus on what should be done, the "thou shalt" rather than the "thou shalt not." But we knew that we had a serious problem when we first tested the SciJourn approach in classrooms and received many, many plagiarized articles. Not only is plagiarism an ethical issue, it also meant that students were simply cutting and pasting information they didn't comprehend. In effect, they were learning little. Moreover, because our goal was to publish stories on the web, we ran the risk of very publicly "stealing" other writers' work.

Although there is a lot of buzz about bad journalistic practices, journalists do work under ethical guidelines and the vast majority attend to these standards. Credible news services constantly update their standards, as the internet and world events present new challenges. Probably the best set of ethical standards, which is available online, has been developed by the Society of Professional Journalists (*www.spj.org/ethicscode.asp*). However, these standards are too complicated and too long to be of much use to students in a high school science class.

To meet the needs of SciJourn we have boiled our standards into a useful acronym: SLAP—no stereotyping, no lying, no advertising, and no plagiarism. Anyone who violates the rules will have their story "SLAPped" down. It is a wonderful mnemonic because it can be rearranged to: Climb the ALPS; swim the LAPS, let's be PALS. Choose the one you like best. Andrew Goodin, a SciJourn teacher, wanted a more positive acronym and created TUNA: tell the **T**ruth, be **U**nbiased, be **N**eutral, and **A**ttribute your sources. The success of the acronym will come not in the mnemonic chosen, but rather in the explanation and follow-through.

In the course of editing hundreds of stories, we have seen all these ethics violated. Laura Pearce talks with teens about the nature of their offenses: a capital crime is a violation of the SLAP rules while other offenses are best compared to civil infractions and minor misdemeanors. Violating SLAP is a capital offense. And our editor tells students outright that if one of his journalists plagiarized, they would be fired immediately ... no ifs, ands, or buts. Laura tells them that they are lucky that they met her before the editor!

We introduce the ethics issues as soon as students begin thinking about writing. Plagiarism, the problem most teachers worry about, is best addressed if students know that they will be revising their stories. A simple note from the teacher, "please reword," alerts the author that you know this isn't their language. As a rule, if the sentence is really well written or full of jargon, we run through a search engine to see if it has been copied. Sometimes it finds a plagiarized section, but we also run into students who are simply talented writers.

A confusing caveat: press releases are technically OK to plagiarize. Often press release language will appear word-for-word on dozens of websites, a sad commentary on what passes for news on the web. For the sake of helping students to understand the nature of plagiarism, it can be valuable to insist that this type of information be attributed to a press release or the organization promoting the work.

Stereotyping often emerges as editorializing or as one-sided arguments. In some cases, the teacher only needs to prod the student to get back on track; other times you are up against some powerful beliefs and a closed mind. For example, we have run into a small group of students (and their parents) who distrust western medicine. To them *naturalnews.com*, a site devoted to alternative medicines, has the same credibility as the National Institutes of Health. It can be a challenge to get them to acknowledge the limitations of alternative medicines. (Similarly, a good story involving western medicine should always acknowledge side effects and lack of effectiveness for some.)

There is no excuse for a story being an advertisement. However, sometimes, a good topic seems to walk a narrow line between ad and science. A story on "green bags"—bags that claim to retard fruit spoilage—posed that problem, especially since the author's family was a fan of them. Much of the initial story's information came from product manufacturers and was presented in the most positive light. To make the story publishable, the author found a consumer report critiquing the product claims and kept the language neutral.

Lying is harder to ferret out. Sometimes, it is the student making stuff up to fill out details, rather than doing the hard work of research. If a section doesn't ring true, then maybe it is not. Talk to the student. In a conversation you can learn more than by writing a note on the paper.

Using the SciJourn Standards

We live in a standards-driven school environment and for that reason alone it probably makes sense for us to be clear about our SciJourn standards, but there really is more to it. First, these standards are designed to help you think about science literacy and science journalism before even embarking on a SciJourn project. Many teachers have described the standards as putting into words the ideas and concepts they are already trying to teach. The standards gave them a new language. These same teachers also talk about reading and viewing science news with a new perspective, one they feel called upon to bring to their students.

The standards can next be used to guide your instruction. They are designed to help you set priorities, prepare lessons, and decide what to teach and when. We do *not* suggest handing the students a list of these standards on the first day of class. Instead, we recommend putting the standards into use through modeling as discussed in the next chapter, attending to standards-related issues as they arise in the articles you are sharing in your read-alouds and think-alouds. Students will gradually begin to recognize the standards for themselves, even before they are formally presented.

Figure 4.3

SciJourn Standards Translated Into Student-Speak.

Standard	Student-Speak Version
A good *SciJourner* student article ...	
I. Has most or all of these elements: local, narrow, focused, timely, and presents a unique angle.	Why am I interested in this topic? Does it relate to something in this community? Is it something I can write about in 500 words? Is my story timely, unusual, quirky, or offbeat?
a. Findings are meaningfully applied to personal or civic issues.	Will my readers care? What's the personal or local impact?
b. Readers' likely questions are anticipated and addressed.	What might my readers want to know about this topic? Are there "gaps" in my story that could leave my readers confused?
II. Uses information from relevant, credible sources including the Internet and interviews.	Where can I find good information about my topic? How do I decide if the information is good?
a. Successful authors use internet search terms and search engines effectively.	What words should I type into Google or another search engine? Are there better terms to use for the search? How will I find the right words?
b. Successful authors privilege data from credible government and nonprofit sites and can justify the use of "other" sites."	What are the "go to," super-reliable websites that I need to check on this topic, i.e., which government agencies or nonprofits can provide the information I need? Given my topic, what other websites would be good to use to represent other opinions and information? Can I trust these sites?
c. Successful authors are able to locate and query experts and relevant stakeholders.	Where might I find an expert to interview on this topic? Who has been directly involved in my issue? Can I get access to them?
III. Is based on multiple, credible, attributed sources.	Who says? Why should I trust them? Who can confirm this? Who has a different opinion?
a. Sources are relevant and reliable.	What makes them an expert on my topic?
b. Stakeholders with varying expertise and experience are consulted.	Have I covered all the "sides" to this story? Is there anyone else I should hear from? Who benefits and who loses; do I represent important stakeholders?
c. Sources are identified and basis of expertise is explained.	Do I give readers enough information to decide for themselves if the information is trustworthy?
d. All assertions, numbers, details, and opinions are attributed.	Are sources clearly identified? Could my reader check me out or get more information if they want it?

Continued

Figure 4.3. SciJourn Standards Translated Into Student-Speak *(Continued)*

Standard	Student-Speak Version
IV. Contextualizes information.	What's the background on this topic? Why does this topic matter? How many people are affected? How is this better than other technologies or ideas? Why are scientists interested in this topic? Is this cheaper than current technologies?
a. The article tells why the information presented is important.	Why should the reader care? How does this story relate to the wider world of policy politics, health, economics, ethics, etc.?
b. The article indicates which ideas are widely accepted and which are preliminary.	Is this a new science or technology idea that is not widely accepted or is this really not surprising to people in the field?
V. Is factually accurate and forefronts important information.	Do I understand what I am talking about? Did I check my facts? Have I answered who, what, where, when, and why in the opening paragraphs?
a. Science connection is evident.	Where's the science?
b. Difficult concepts are explained.	Will my readers understand what I am saying? Did I say how the discovery came about?
c. Precise language is employed.	Will a picky person be able to pick away at what I am saying? Is my language clear? Am I using science words correctly? Do I define them for my readers or help them create a mental picture?
d. Quantitative measures are given in correct and comparable units.	Have I included the numbers that help me make my point? When the numbers are very large or small, have I included a comparison to help readers picture the number?
e. Information is up-to-date.	Is there anything newer on this topic?
f. Captions and graphics are checked for accuracy.	Does the picture and caption help tell my story or give the reader a clearer picture of what I mean? Are all pictures and graphics related to the story? Are graphs easy-to-read and correctly labeled? Are photos copyright-free or do I have permission to use them?

At this point—as the students are writing—it does make sense to bring the standards out in some wholesale fashion. As students write and revise, you will want to be able to remind them of what counts as good work. It may be helpful to refer to the "student-speak" standards in Figure 4.3 or create your own version of the standards with your class in language and with examples your students find more accessible and memorable.

Figure 4.4

An Annotated Article Showing SciJourn Standards

RING! The school dismissal bell sounds, forcing you out of your nap and back into consciousness. This scenario is one that has been familiar to students for years, and various researchers from around the world are now crediting carbon dioxide as one factor forcing students into a slumber during class time.

> **Comment:** Standard I: The topic of this article—carbon dioxide in school classrooms leading to drowsiness—is narrow, focused, unusual, and of interest to a teen audience.

Among the researchers studying carbon dioxide's effect in classrooms are David Sundersingh, who works on the design, architecture, and engineering of eco-friendly buildings with the national firm Fanning/Howey Associates, Inc., and David Bearg, an independent mechanical engineer. Their 2003 article on indoor air quality names carbon dioxide as one of the pollutants widely prevalent in the classroom.

> **Comment:** Standard V: The science connection of this article is immediately apparent.

> **Comment:** Standards II and III: This is the first attributed source for this article. The qualifications of the two researchers are explained, however the attribution would be stronger if the writer had included the name of the publication where the article originally appeared.

Carbon dioxide is released in all combustion and human metabolic processes such as respiration. In their article, authors Sundersingh and Bearg state, "In the developed world, 90% or more of our lives are dependent on the indoor air quality in homes, workplaces, and vehicles."

> **Comment:** Standard IV: The information here and in the paragraph that follows establishes the context of this article by answering the question: why should we care about indoor air quality?

Despite air quality's significance to one's well-being, researchers are finding troubling evidence of poor air quality in schools as a result of weak ventilation, inadequate filtration, and insufficient maintenance of air-managing equipment. According to the Environmental Protection Agency, many schools currently have air ventilation rates that are below recommended levels, leading to high levels of carbon dioxide in the classroom.

> **Comment:** Standards II and III: The second attributed source, this time to a respected government agency.

Pamela Becker Weilitz, a doctor of nursing practice and a pulmonary nurse specialist at the John Cochran Division of the St. Louis Veterans Affairs Medical Center, agrees that human respiration leads to increased levels of carbon dioxide in a room. However, she also claims, "Other factors such as the ventilation in the room, the number of people in the room and the temperature of the room also play a role in carbon dioxide levels."

> **Comment:** Standards II and III: A third attributed source to a third kind of stakeholder, a health care expert. The author interviewed this source herself, which could have been clarified in the text of this article through a phrase like "she tells SciJourner."

Hence, when a large group of students reside in a classroom with little air ventilation for an extended period of time, the surrounding air is apt to have increased carbon dioxide levels as a result of the inevitable process of exhaling, which cannot be limited, and poor ventilation in the room, which can be improved.

A 2008 study by scientists from University College London (UCL) and Reading University (both in the U.K.) showed that this increase in carbon dioxide as the school day progresses actually causes students to become drowsy. To discover this, the researchers required students to answer a health symptom questionnaire and take a test that measures their ability to concentrate on schoolwork. The researchers found that students in classrooms with high levels of carbon dioxide and poor ventilation demonstrated an increase in health symptoms and lower concentration scores. They also compared ventilation in 50 year-old school buildings and new, heat-trapping school buildings, and found a correlation between carbon dioxide exposure and feelings of lethargy.

> **Comment:** Standard V: The date establishes that the work is recent.

> **Comment:** Standards II and III: Because these universities may not be familiar to readers of SciJourner, the author included their location. This fourth source for the article is to a study that connects air quality directly to classrooms and learning.

> **Comment:** Standard V: The author of this article is able to summarize this study in just a few sentences. To do so, she chose the aspects of the study that were most important to her article and translated them into language that her readers could understand without distorting the scientific content.

The ventilation rates in both old and new buildings "were equally appalling," says UCL researcher Dejan Mumovic, who claims carbon dioxide levels in stuffy classrooms have a major effect on student learning.

Although researchers believe carbon dioxide dooms students to feelings of sleepiness, have students and teachers recognized increased drowsiness during the school day's progression? To answer this question, a survey was distributed to 85 high school freshmen and juniors, as well as 23 teachers, at a local, west county high school in Missouri that was built in 1954. Because the school is over 50 years old, it could have poor air ventilation in its classrooms.

> **Comment:** Standards I and II: The author makes the story local through the use of a survey, a fifth source of information.

According to the survey, over 50% of teachers usually keep their classroom doors open, but over 80% do not open windows often or ever, which could limit the classroom's air ventilation. Although the teachers as a whole did not think their students become more tired throughout the day, over 85% of the surveyed students admitted to feeling at least somewhat more fatigued as the day continues.

> **Comment:** Standard V: Because the survey had a large number of respondents, percentages are appropriate.

We understand the tendency of educators to want to turn the standards into scoring guides. Please note that several teachers and we on the research team have tried putting these standards into rubrics and every effort to date has been terribly unsuccessful. Please restrain yourself until you have read about our alternatives to the rubric—a science article filtering system and a calibrated system of noting poor, medium, and good attempts to employ the standards (Chapter 10).

Figure 4.4 presents an annotated example of a published student article so you can see how all the standards work together. Additional annotated examples are available on *www.Teach4SciJourn.org*. As you read over them and notice how different each version of "success" is, you'll begin to appreciate the reason rubrics have failed.

We realize that these standards are, in effect, a hybrid, born out of the science journalist's perspective and commitments, but tailored for use in the classroom. These standards are designed to be used by teachers to help students become better readers and information gatherers, to focus their thinking, and to help them become science-savvy citizens. Remember, the SciJourn standards and principles are aspirational; we don't expect every story to have it "all." Even the best newspaper articles miss pieces, either through oversight or design. Even a very good article may not do all of the things we mention in the standards equally well. The editor of our newsmagazine, *SciJourner*, uses these standards as a gatekeeper and continues to ask for revisions based on these standards if a student wishes to get published.

As with all good standards, we view these goals as "in development." They have gone through what seems like endless revision and yet we make no promises about their status as complete. Please check into the website *www.SciJourn. org* to download the latest version for your students. And do let us know if you have any suggestions for revision.

References

Centeno, I. June 3, 2010. Why airtight? *www.SciJourner.org*

Distler, A. June 4, 2011. Is bottled water safer than tap water? *www.SciJourner.org*

Emrick, D. July 20, 2010. The secret number one. *www.SciJourner.org*

Goebel, H. April 23, 2010. Horses and healing. *www.SciJourner.org*

Harris, J. July 16, 2011. Glaucoma. *www.SciJourner.org*

Jacobsmeyer, L. May 19, 2011. *Sunscreen—Cancer preventative or promoter?* *www.SciJourner.org*

Johnson, D. July 6, 2011. *Michael Jackson: Heart attack or cardiac arrest? www. SciJourner.org*

Northwest Regional Educational Laboratory. n.d. 6+1 Trait Rubrics. *http:// educationnorthwest.org/resource/464*

Pool, R. September 3, 2010. The heartbreak of Potter's syndrome. *www.SciJourner.org*

Ragland, N. July 27, 2010. *What's wrong with my head? www.SciJourner.org*

Stavri, A. July 21, 2009. Thinking about a tattoo? *www.SciJourner.org*

Turner, K. November 7, 2009. Emergency medicine from a teen's view. *www. SciJourner.org*

Turner, K. April 13, 2010. The case of the mysterious allergic reaction. *www. SciJourner.org*

Whistler-Brown, W. April 15, 2010. Video games and science. *www.SciJourner.org*

CHAPTER 5

SETTING THE STAGE BY MODELING

Several students were aghast. Many others were intrigued.

"You mean I washed my face in germy water this morning?" one asked.

"Can the microbes in my shower make me sick?" asked another.

"Well," responded their biology teacher, Ms. Barker. "I don't know. Let's read on."

Ms. Barker was reading an article with the headline *Dangerous Pathogens Live in Showerheads* that she had found at *www.news.discovery.com*, a science news website associated with the Discovery Channel. She went on to read aloud to the class that cleaning the showerhead with bleach doesn't help and that the spray nozzle creates a fine mist of water droplets that can be inhaled deep into the lungs while showering. Those water droplets contain mycobacteria.

Ms. Barker stopped and pondered for a moment. "I'm also wondering if this is a health concern. Have any of you heard of mycobacteria before?" Ms. Barker was genuinely interested, her curiosity piqued. "I don't know, and you surely recall that not all bacteria are harmful," she continued. "Shall we read more?"

Although the bell was about to ring signaling the end of class, it was obvious that most of the students wanted to hear more. The article was of interest not only because this was a biology class and the students had acquired a degree of familiarity with the topic, but also because of the personal connection—students shower, and they do so often.

"Oops, out of time," Ms. Barker said. "We'll continue tomorrow, but if you're really interested, try searching 'pathogens in showers' tonight."

There was a strategy in her methods.

This was not the first time Ms. Barker had read aloud to her class, nor was this experience merely a "read-aloud." Both Ms. Barker and her students were also engaged in a *think-aloud*—pondering the content and posing questions about its relevance. Together they were reading about and considering findings recently released by a major university. It was up-to-date, it was credible, and it was a topic that connected with the students.

Ms. Barker and numerous other high school teachers have discovered the power of reading aloud to high school students. These teachers have found that students of any age benefit by listening to and thinking about articles on a range of topics. Many of us remember story time during our elementary school years. Studies have identified the positive effects of reading and thinking aloud to children (VanDeWeghe 2004). Through read-alouds, children encounter advanced vocabulary as their teachers model ways readers approach and respond to text. In this chapter we explore similar benefits for high school students.

Making the Connections

Literacy professionals often think about the connection between reading and writing. How does reading impact the ability to write and think? What do competent writers notice when they read? Interestingly, when we began this project we tended to be product-oriented and focused almost exclusively on how best to help students create good articles. Then we met up with teacher Mike Ruby. Mike is a really good writer, but when it came to giving time over to student writing, he was hard to budge. He was happy to model his own reading strategies, attend to content, and talk about how a news article worked. And since we were not developing a how-to curriculum, we cheered Mike on and watched.

Mike experimented. With one class he simply took 5–10 minutes at the beginning of a period to go over an article he found interesting. The article may have been one that seemed to be relevant to what he was teaching in his physical science class or perhaps it was an article that demonstrated some principle of journalism he wanted them to better understand. Sometimes they were articles that related to another article he had already read or a piece about a local or current event. Meanwhile, read-alouds/think-alouds (RATAs) were absent in Mike's other class.

Although his sample was not large enough nor the experimental design sound enough to claim this approach as scientific, Mike found that the kids in his "experimental" class did no better or worse on the end-of-course exams than the students in his "control" class. However, the kids in the "experimental" class seemed to like Mike better since they were really getting to know him through his thinking aloud as he read with and to them. During a think-aloud, Mike revealed aspects of his personality, likes and dislikes, beliefs, and values. As students came to know their teacher, a heightened rapport was established. So, Mike ended up liking the kids he read to better as well, so much so that he gave up his experiment and began reading to his "control" group as well. Through the read-aloud process he became better acquainted with and connected to the students. That rapport actually enhanced the learning taking place throughout the year.

"The easiest place to begin is with a read-aloud. It's low risk, high interest, and high impact."
—SciJourn teacher

What the SciJourn program realized from Mike's "experiment" was that a RATA takes very little time away from the regular class activities and has the amazing potential to help young people learn what sophisticated readers do when they encounter text. In fact, if we are called to do a professional development series for a school we *always* begin with the RATA. It doesn't negatively impact the scheduled curriculum and requires very little planning. Students respond positively (unless the teacher turns the activity into round-robin reading—an unmitigated disaster!) and the positive feedback from students inspires teachers to do more.

Too often the approaches adults take to "making meaning" as they read and write are not in any way obvious to students. Even when the competent reading teacher recognizes this problem and tries to specify generalizable strategies (e.g., visualize when you read, stop when you don't understand and go back to the beginning of that paragraph), she doesn't know enough science to provide context-specific hints and connections that promote science literacy.

Of course, we, as educated adults, can take many different journeys or paths through a book, article, or news story. So how might a teacher decide which paths to pursue in a RATA? One good strategy, the one used by Ms. Barker in our opening example, is to simply stop and curiously reflect on the text in front of you, indicating what connections you have made and perhaps asking what connections the students are making.

In the pages that follow we suggest a somewhat more systematic approach, encouraging teachers to consciously choose, name, and track the kinds of things they are modeling. Remember: It is the "think-aloud" component that sets the read-aloud apart from just orally sharing an article with a class. It is the connections to the article and topic, to the curriculum, and to one another that the think-aloud provides. The think-aloud is what makes this activity an effective means of enhancing science literacy.

In order to get started, we have listed a number of paths you may wish to embark upon as a reader; each is tied to particular aspects of science journalism and science literacy. Please note: It is impossible and not even vaguely desirable to try to include all or many of the strategies listed below in a single RATA. The targeted, purposeful RATA is your goal. Always remember, it is you, the teacher, who needs to think carefully about what aspects of science literacy you seek to highlight in your RATAs, and it is your responsibility to keep track of the purposes of the RATAs you have modeled.

To help in that task, consider creating a template to be put online or on a bulletin board to remind students of these "paths." If you link to the article or photocopy it for the bulletin board, add a heading like this: "Remember how we used this article to think about...?" and then you or have a student fill in the blank with the things you touched on.

A Quick "How To"

- Keep it brief. You don't even need to read the entire article.

- Ideally, show the article on the screen as it is being read. This is an unusual opportunity for students to both see and hear text.

- Make it interactive. Invite student comments and reflections.

- Ask about the credibility of an article. Who believes the article is accurate? What are the clues?

- Point out connections you notice during the read-aloud—connections with other texts, with personal experiences, with other topics, with local issues.

- Examine issues related to the SciJourn standards:

  o Are the sources credible? How many viewpoints are represented?

  o Are all the sources fully attributed? What do you learn from the attributions?

  o Relevance—who cares?

  o Is it factually accurate and up-to-date? How can you tell?

- Examine the way science works. Read stories about contemporary discoveries and the scientists who have made them.

- Invite skepticism. Push for it during the read-aloud of a poorly written article.

- If students are interested, follow the hyperlinks in an article.

- Read-alouds are especially effective if they are entertaining. This is a great time for the teacher to perform.

- Sometimes offer choices—two or three headlines from which the students select the one of greatest interest. Our guess is that even some of the nonwinners will be read if you make them available.

- Invite students to select articles to read to the class themselves—this is a great opportunity to assess what is selected and how the student reader interacts with it.

- Point out the image, figure, and table that accompany the story. How does it help the story? Is it the best illustration for that article?

Thinking Aloud

Because RATAs are short activities that serve to model the habits of mind of a scientifically literate person, they can be done any time you have a few moments to spare. For instance, before the bell rings or while a couple of students are finishing up their work, have a cache of articles ready so that you can pull them up anytime. *SciJourner.org* is also a useful resource.

Read-alouds/think-alouds also prepare students for writing their own articles by introducing them to a new writing genre, highlighting concepts that you will return to as they begin their own work and developing a database of examples that you can refer to in your own responses to their work. In order to evaluate your own teaching efforts as they relate to the read-aloud, ask these questions:

- How did this RATA help students to become more scientifically literate?

- Will it offer them the kind of skills they will need in order to evaluate science information 15 years out of high school?

- Has it helped them analyze and evaluate information and sources?

Figure 5.1

A Few Sources of Quality Read-Aloud Materials

www.scijourner.org

www.sciencenews.org

www.sciencenewsforkids.org

www.ScientificAmerican.com

www.nytimes.com/pages/science

www.newscientist.com

www.news.discovery.com

www.bbc.com

Ready ... Set ... RATA

Figure 5.2 (pp. 60–61) is an article from *SciJourner.org* written by a high school journalist. For teachers unfamiliar with the read-aloud/think-aloud process, we've provided scripted comments in the margins to serve as a model for you. The direction this RATA takes depends, of course, on student comments and questions and those will differ from class to class. This read-aloud is merely a start. You can even think of it as a drama with you as lead!

Figure 5.2

An Annotated Article for a Read-Aloud/Think-Aloud

How Green is Your School?

Written by Mobeen Mian, Francis Howell High School

> **Comment:** This article was written by a high school student in Missouri.

Ever wondered how much trash comes out of a school? Has your school ever gotten a waste audit a method of estimating the total amount of waste discarded, the cost, and the amounts of waste that can be recycled or reused?

> **Comment:** I wonder if our school has...have any of you heard about this happening at our school?

Schools are a huge producer of waste. Can schools be more ecofriendly?

> **Comment:** Did that lede capture your attention? Do you have a good sense of what the story is about?

According to the website, *Earth Share*, an environmental group dedicated to promoting environmental education, on average, one student will produce 67 pounds of waste in a school year from bringing lunch to school. This adds up to 18,000 pounds a year for an average elementary school. The majority of the wastes are paper materials and food waste, according to *Recycle Now*, a website that tells you what can be recycled and how to go about doing it.

> **Comment:** Does anyone know anything about this group? We should look them up later on.

> **Comment:** How many of you bring your own lunch to school? It says here that...(continue reading)

> **Comment:** Hmmm...I've never heard of this website—sometime we should visit it and try to decide if we think this is a credible source.

If schools operate in a more environmentally conscious way, not only does it create a much cleaner environment and reduce pollution, but lowers the school's cost of operating. For example, research conducted by the U.S Department of Energy (DOE) shows that one of the highest expenditures for a school is their energy cost. Public schools spend over 8 billion dollars a year in energy costs! School districts can lower their energy cost by up to 20–30% by implementing energy efficient operations and maintenance, such as switching from T12 to T8 fluorescent light tubes; T8 use less energy but give the same amount of lighting, according to DOE.

> **Comment:** Is the Department of Energy a credible source of information? How do you know?

> **Comment:** This is interesting to me, but I wonder about the costs. Are these lights more expensive? I wonder if we have these kinds of lights in our building. How could we find out? Are there any questions you have about this so far? Okay, let's read more.

Another great way to save energy is simply turning off unused computers, which cost $0.01 to $0.03 per hour to leave on, according to DOE. The best way would be to educate the school staff and students on energy consumption and ways to save energy so staff can implement this energy plans adds DOE.

> **Comment:** Okay, so that's what, about $.78 per day—that can really add up, especially with a lot of computers. Are there any advantages to leaving the computers turned on? There might be...

The staff of Francis Howell High School in St. Charles, MO, has taken the initiative to decrease the school's effect on the environment. "Our school is working with a new waste and recycling company that recycles everything but food. The year before the school used to throw everything in the trash with the old company," says Ann Clynes, a lunchroom employee. "This new company gives the school a 100% increase in waste being recycled along with eco friendly detergents," adds Bill Clynes.

> **Comment:** I wonder who Bill Clynes is. I wonder if he's related to Ann, the person cited above.

The Clynes also mentioned that the school had employed reusable plastic red trays to substitute the paper trays to reduce cost. Unfortunately, students wouldn't return the trays and threw them away and we had to switch back to the paper trays.

Most of the student body believes the school could be doing a much better job.

Continued

Figure 5.2. An Annotated Article for a Read-Aloud/Think-Aloud (*Continued*)

"The school is extremely inefficient in energy consumption such as lights and computers," says Brendan Gowen, a senior at Francis Howell High. "The school leaves most of the lights on at night when they really don't have any reason to do so."

> **Comment:** Are the lights here on at night? Have any of you driven by?

Zach Shultz, a senior at Francis Howell High explains, "The school is not doing enough to create a more eco friendly facility and the last thing on the school board's mind is the efficiency rating of the school, the school can do much more." Other schools are becoming more eco-friendly.

> **Comment:** This sounds interesting—maybe one of you would like to do some research about our own school and possibly write a similar article about us.

"I agree that the school is currently inefficient and not very eco friendly," says School Principal Dr. Chris Greiner. "But the new building that is currently being built [scheduled to be completed 2011] is much more efficient as far as lighting, and HVAC [heating, ventilation and air conditioning] with innovative features. The classrooms will have motion detection lighting which will sense if anyone is in the room and turn off the lights when the room is not in use. The school district is doing a great job on making the protection of the environment and the efficiency of their facilities a priority; like how the school will be expanding its cell membrane water treatment plant," says Greiner. A cell membrane water treatment plant is an environmentally-friendly procedure to treat water before it flows into creeks.

> **Comment:** How many of you would have the courage to talk to the principal about something like this?

> **Comment:** This all sounds great, but I still wonder about the costs…

Colleges are also coming under scrutiny. The Sierra Club's October 2010 100 Coolest School Survey, which compares conservation and energy efficiencies of universities and colleges nationwide, lists Washington University in St. Louis at 43rd and University of Missouri-Kansas City came in 95th.

> **Comment:** Is this information about the Coolest School important to include?

A similar challenge implemented on K-12 schools, called the Green Cup Challenge, was established by the Green Schools Alliance. They challenged schools to lower their carbon footprint by making simple changes like dimming the lights and adjusting the thermostats. With 161 schools participating, they say that they were able to achieve the equivalent of taking 162 cars off the road or planting 764 trees!

> **Comment:** So what do you think? What questions do you still have that were not answered? Does this all sound credible to you?

What's Next?

After completing the RATA, it is important to restate the kinds of questions that you asked. For example, "In this story we asked questions about relevance, does this story match your own experience? What about credibility? Did you find all those sources believable or useful? How about factual accuracy? What were the real costs?"

The RATA in Figure 5.2 shows how the various paths meander through the story and offers one possible approach. Remember, many stories do not need to be read in their entirety, especially if time is limited. Also, the questions posed at the end of the read-aloud could just as easily be asked earlier in the reading. With any RATA, flexibility is key. Time available and the tenor of the class should guide the process. There are no absolute right or wrong ways to share readings with students.

Press Releases Versus News Articles

Materials for read-alouds fall broadly into two categories—press releases and journalistic articles. Press releases are typically issued by institutions, corporations, or government agencies to announce new or especially exciting discoveries, technologies, and procedures. Often they highlight new data derived from one or more research studies. Basically, a press release is designed to say: "This is who we are and here is what we found." Press releases are an important means of communication, sent to journalists in hopes that a news story will be written based upon information provided.

The credibility of a press release depends upon the source. Generally, government agencies and education institutions promote information that is credible. They are often well-written and carefully vetted before publication. Yet, they are also one-sided and tend to hype their news as more broadly accepted or successful than it might be. They are very useful however, in figuring out what's new. At their worst, a press release promotes a product or false information.

When we first began this project, we thought the difference between a press release and a news article was important enough to warrant spending a lot of time and energy on. However, as we learned more about how SciJourn works in classrooms, we realized that the distinction between a news article and a press release from a credible source was extremely subtle. In fact, the distinction is not even well understood by scientists working in the institutions that report on their research, since the press release is often picked up by a legitimate news source.

Many science, technology, and medical news stories begin with a press release or press event. Reporters treat press releases and events as a single, credible source. To complete the story, they look for other credible sources for comment, verification and context. However, on some websites what is called a news story is really just a press release; few if any changes are apparent.

In most classes, you probably won't want to spend a lot of time comparing news articles to press releases (particularly press releases from credible sources). But we do believe it is worth a brief discussion—students often are familiar with press releases put out by celebrities or sports figures. Press releases are also a good source for thinking of possible article topics.

To demonstrate how a press release fits into the landscape of science information, try this:

- Pull up a press release on the internet (science-related press releases can be found at *www.eurekalert.org* or *www.sciencedaily.com*).

- Copy the first few sentences.

- Paste into Google or another search engine and see what results come up.

Because institutions issuing press releases want them to be used by media outlets, they don't have to be cited; copying them word for word is not technically plagiarism. Searching the first few sentences of a press release, especially press releases on health or technology, will usually yield several hits to other websites where the press release appears with the same wording. Demonstrating this for students can open up interesting conversations about press releases and the ethics of journalism.

For students ready to understand the nuances of journalism, more detailed lessons on the differences between press releases and news articles may be worthwhile. These lessons, found in our online supplementary materials, ask students to focus on perspectives missing from a press release. As such, these lessons help students critique the discoveries being described in a press release.

Lesson Ideas

As Mike Ruby demonstrated at the start of this chapter, if the only thing related to science journalism you ever did in class was the RATA, we believe you and your students would benefit. Variation in approach is what will make your RATAs interesting. Let's take the notion of credibility. Getting students to think about issues related to credibility is obviously important. How do students decide what to believe? (Lots more about teaching credibility in Chapter 7: Finding and Keeping Track of Sources and Chapter 8: Original Reporting: Interviews and Surveys, but these think-alouds lay the foundation). As a beginning, get students to think with a partner or with the whole class about whether they believe the information in an article and why. Then discuss what clues you, the teacher, have used to decide. This is an opportunity to present credibility as a continuum, not a black-and-white issue—all sources in an article probably have some kind of expertise, but they all also have bias. How do you decide

Photograph 5.1

Teacher Sam Berendzen, wearing a colorful lab coat, leads a read-aloud in his biology class.

Photo: Laura Pearce

what weight to give information from different sources? As students become more comfortable thinking about credibility, you can ask for volunteers to think aloud about the credibility of sources during a RATA.

The variations for RATA are limitless. Here are a few more ideas. Note: More detailed lesson plans for RATAs are available online.

- *The curricular connection.* Tie your read-alouds to what you are studying. The content of the article can remain in the background while you introduce another aspect, e.g., credibility, or it can be foregrounded, e.g., "We were studying helium last week. Look at this novel use of it."

- *"Isn't this interesting!"* Cool or surprising science and technology stories are always popular.

- *The daily news briefing.* Rather than a single article, display a website with several science articles on a single page. Scroll through the headlines, examine the ledes, read further if students seem especially interested.

- *Look for multiple sources.* If you had to ask someone about this topic, who would it be? How well did the journalist do?

- *Looking at what's new!* This is an opportunity to read about the work of scientists and the discoveries they have recently made and to bring in current events, such as a recent flood.

- *How science works.* Stories about research can offer opportunities to ask about or scrutinize research methods and to examine the logic of the data presented, noting what's missing or what "as a scientist" you would like to know more about.

- *What do we mean by quality?* Invite skepticism, especially during the read-aloud of a poorly researched article. Help students differentiate between what they might call "high quality" articles and others. What do we mean by quality? Chapter 2: Science Journalism Goes to School articulates our standards for an excellent news story, but remember these standards are aspirational. That means there is a quality continuum for news stories, ranging from the awful and unbelievable to the thoughtful and engaging. It can be fun to throw in a poorly written article to see what comments and questions emerge from the students. This can serve as yet another means of informal, formative assessment.

- *Ask the tough questions.* Invite the students to ask the tough questions journalists ask about a topic or article, especially when reading a press release to the class. In Chapter 8: Original Reporting: Interviews and Surveys, we present an exercise in which students come up with the

questions first, then test themselves against a professional podcast or videocast news story.

- *How close are they?* This is an idea for particularly advanced students or those you have for a long period of time. Once students understand how to recognize sources and determine credibility, you can begin thinking about the kinds of sources included in an article. Are they all closely connected or do they really represent different points of view? Examine the relationship among sources of information according to their financial, professional, and personal connections.

- *Captions and photographs.* Examine the photographs and the captions. Ask how important the artwork in an article was in creating interest. Look at the labels on graphs and charts—or remove the labels and see if students could guess what those labels might be.

- *Text-to-text connections.* Point out the connections you notice with other stories you have read, with lab demonstrations they have done, with textbook information and grade-relevant expectations that have been previously noted.

- *Personal experiences of the reader.* Connect with your own personal experiences or those of others in the class. Try to guess who in this class might be interested in this article and see if you are right.

- *Text-to-world connections.* Connect your reading to local issues. A newly named Superfund site? An unusual weather occurrence? Is it time to upgrade school computers?

- *The follow through read-aloud.* Click on some hyperlinks of relevance. Talk about the sources you identify.

- *The "I don't get it" read-aloud.* Demonstrate for students what you do when you encounter difficult vocabulary or concepts. They rarely learn to figure their way out of conceptual trouble in American schools.

- *Can you second-guess me?* Ask students to jump in and do the "thinking" for you. Stop where you would normally stop and see what kinds of questions and comments the students ask.

- *Who says?* To focus student attention on the importance of attributing all information that is not commonly accepted, take an article and remove all of the attributions to sources. Ask students to call out or raise their hands every time they want to know "who says" a particular piece of information. When students ask the question, sometimes provide an incomplete attribution ("Jane Doe" or "a government agency") and encourage them to ask follow-up questions.

- *Looking at form.* News articles are a genre unto themselves. Here is a way to introduce the notion of the inverted triangle, the nutgraf, and the 5 Ws. When you do read-alouds for form, you may want to look at a single characteristic—the lede, for example—across several different articles. See Chapter 2 for more information about journalistic form.

- *The "expert" RATA.* Invite a scientist into the classroom to do a read-aloud. If you have access to scientists with different areas of expertise, invite several in throughout the year. Seeing how they read material both within and outside their area of expertise can be revealing for students. You might also consider calling a scientist by phone or video chat on the internet to offer their read-aloud.

- *Reading graphs and infographics.* Find an article with a graphic; alternately, find a stand-alone infographic (*www.good.is/infographics* is a great source). Demonstrate how you read these challenging multimodal texts.

- *Visualization.* Using an article with strong visual imagery, encourage students to close their eyes and picture the words as you read. Stop and ask students to describe what they see. This is a good strategy for reading all kinds of text.

- *Which article would you like to read and why?* Show the students several headlines or ledes and ask which they would like to hear.

The Best Idea Yet

The very best modeling you as the teacher can provide is to work on your own article with the students. In fact, if you can work ahead of your students by a day or two, they will have a keen idea of what you are looking for and the problems they are likely to encounter. Think aloud about the problems you face—why you drop ideas or latch on to others; struggle to come up with a topic; search for multiple, credible sources; are challenged to write and how you deal with revision. The remainder of this book is designed to help with this writing process.

Reference

VanDeWeghe, R. 2004. Reading apprenticeships. *English Journal* 93 (5): 90–94.

CHAPTER 6

"WHAT'S YOUR ANGLE?"

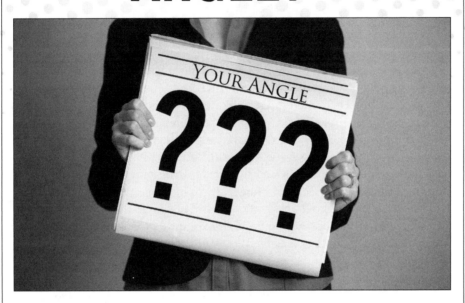

The best curricular and classroom practices provide a window into, and practice with, activities that go on in the real world. But when these activities make their way into the classroom they change. In the real world less time and support is provided for novices and in the classroom more attention is devoted to feedback and supportive critique. Because of these differences we no longer call our journalism an "authentic" activity, but rather a "hybrid practice."

In the authentic world of the newsroom, the process of finding a story topic goes something like this: Either the reporter "pitches" an idea to the editorial team and it is discussed or, alternatively, an editor assigns a reporter to a story. The editor might also ask a writer to investigate and research to find out if there is really a story there. The decision on what gets into the paper, or on air, lies with the editor. The job of the editor is to serve as a proxy for the reader, viewer, or listener. The editor attends carefully to staff reporters and makes decisions about which avenues the reporters are able to pursue and which should be pursued.

Our school version of the newsroom is built on this same model but has been modified for classroom use. There are various ways for students to experience the sharing of story ideas through the pitch, but in all cases the teacher serves, at least temporarily, as editor. As editor she needs to keep certain concerns front and center as she helps students find a story that is:

- doable in the time allotted—i.e., not too broad and not too narrow,

- supported by research that students can grasp,

- personally meaningful,

- of interest to others, and

- compatible with her and the school's culture and teaching goals.

Side note: There is a difference between what journalists and teachers mean by the word *topic*. Newspapers and magazines are in the business of judging the topics their readers find interesting. Unlike teachers, they are not in the business of deciding what people should learn or what is important for educated people to know. Folk remedies and alternative medicines often get top billing over the latest findings on monoamine oxidase (MAO) inhibitors, even though these latter compounds are an important class of antidepressant drugs. An important classroom topic such as Newton's three laws of motion or DNA replication to a reporter is just background information, justifiably omitted from the story.

In the SciJourn approach, differing definitions of *topic* (i.e., topic as the subject of a news article that will entice readers and topic as a unit of school curriculum) become complementary. For instance, in an effort to understand the purpose and risks of auto air bags—accidental deployment, toxic fumes, and the mitigation of potentially fatal injuries—SciJourn writers need to understand and apply those textbook topics heretofore seemingly unconnected to real people's lives.

Size

Journalists have taught us that finding a topic that is "the right size" really matters when trying to write a successful article; that is why SciJourn standard I states that articles should be local, narrow, focused, timely, and present a unique angle (see Chapter 4). Students and teachers alike seem to drift toward big issues, such as global warming or stem cell research. We understand why. These are the topics that are "in the news," and the commonsense assumption is that if the topic is big enough, almost anything will fit. Perhaps surprisingly, the big umbrella approach gets students and teachers into lots of trouble. Big topics tend not to be personal, mask a student's inadequate understanding of a topic, and often result in those loose logical connections that characterize the formulaic five-paragraph essay.

The 500-word (more or less) science news article offers a bite of information that is produced using a limited body of research. As our editor tells students "You can't write about diabetes—that's a book; you can't even write about type 2 diabetes—that's too big, but you can write about how type 2 diabetes affected your grandmother." Explaining global warming in 500 words is impossible; even narrowing the topic to global warming in St. Louis yields mostly clichés.

Figure 6.1

A *SciJourner.org* Published Story

The author was passionate about the topic, kept it narrow and focused, and later in the article discussed the quality of her local drinking water, citing credible sources all along the way.

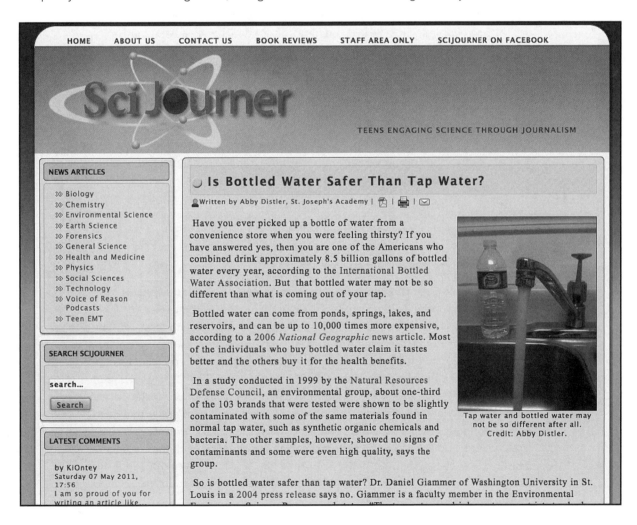

However, 500 words on how global warming might affect the Japanese beetle infestation in your community has real possibilities.

We realize that this advice is different than what students are typically told to do in their other classes or in the debate society where instructors are looking to see if students can argue cogently or develop their own thesis. But the skills we seek to foster in SciJourn have to do with science literacy—helping young people find current, relevant science research about topics of personal or civic interest. We want topics that help build the kind of skills that teens will need later in life. Yes, the big-topic essays may well prepare students for college, but remember that our goal is 15 years from now. All of us, whether or not we attend any postsecondary school, will someday encounter a particular disease we need to deal with. All of us will need to decide which new technology to purchase. Our searches later in life will tend to be narrow, specific, targeted, and timely. Those are the kinds of questions we are trying to help students address now, so that they can practice and build their confidence as decision makers.

Young people seem to be keen on topics about which they have strong opinions. Although we agree that smoking is definitely not a good idea, a story on the evils of cigarettes has all the charm of a bad health class lecture. However, a survey on the smoking habits of local high school students is both doable and interesting. (Chapter 8: Original Reporting: Interviews and Surveys has advice on how to conduct surveys and what to do with them.) Here are the three basic rules for the student:

- Following SciJourn standard I: Keep the topic narrow, focused, local, timely, and interesting.

- Care about the topic.

- Ensure that the topic is "doable."

Grasping the Research

Sometimes a student comes up with a really interesting idea and there is little research to support it or, alternatively, the work is so new that the only available current reports on it are highly technical and truly difficult to understand. The teacher should not be expected to anticipate such problems, but if you do, it is important to let the student know. Topics and the direction an article takes may change as the journalist learns more. Professional journalists drop topics that run into dead ends, and students should be expected to also. For instance, we recently read a press release about a new technology that, using only the video image and gait of a person, can produce an accurate estimate of the person's size, weight, and shape, even a person wearing the baggiest of clothes. Students reading this press release became interested (as was the researcher) in what this might mean for stores where robberies had taken place by trench-coated intruders captured on camera.

But as the research got underway, it turned out the technology was so new that there were few other stories or experts able to comment on the viability of the mathematics undergirding the assumed measurements. Basically the story the *SciJourner* author originally had in mind was not working, but other stories were possible, e.g., interviewing local police about the usefulness of such a technology (in which case the student would have to explain the technology to the police) or looking at video surveillance technology currently in use at banks or supermarkets.

Working off of a press release, newspaper article, or TV segment, in our experience, yields mixed results. In too many cases, the student's tendency is simply to summarize what is already written, not combining or digesting information or putting it together in new and unique ways. We typically push the author to take the next step by getting reactions from other experts or stakeholders, which is what a real reporter does. While we would be surprised if a professor at Harvard answered a query, students have had a good deal of luck getting responses from local experts. See Chapter 8 for several useful strategies.

If a student is fascinated with new research, we suggest that he or she add "something" to a mainstream media story. For example, to augment a story on new research showing that brushing teeth after drinking sports drinks can damage tooth enamel, the author collected data on how many of these drinks were sold in the high school cafeteria along with interviews on whether teens were willing to change behavior based on the findings. In effect, she created her own angle on the press release story by making it local. In sum, if a student is ready to comb press releases for an idea, there is probably something to be found that, with imagination, is worth pursuing.

Teachers also need to be concerned about students picking topics that are frankly too difficult to understand, such as a new discovery about "dark matter," the physics of a curve ball pitch, or how most technologies work. Such topics often lead to plagiarized reporting or long, inscrutable quotes. The best way to deal with this, experience suggests, is simply to ask the student to explain the idea in a way someone less familiar with the topic could understand. It's a move that often leads to a student de-selecting that topic or realizing that he or she needs help with the science basics. The other approach is to find out why the person is interested in the topic, which can, again, lead to a better idea.

We also find "how it works" stories to be problematic. A story on the x-ray detector used at the airport is likely to be boring and read like an encyclopedia entry. In fact, most of the story will likely be copied from Wikipedia or some other content site. What would help is an interesting angle, such as how to defeat the airport scanner or what can the scanner really see.

Let's admit it: because many topics are a "leap in the unknown," the teacher-editor can get it wrong. One frequently pitched topic from members of the ear-bud generation is a story about connections between music and mood. With this in mind, when a student pitched the idea for an article on "how music makes you smarter," we rolled our eyes and sort of groaned. But in fact, this young journalist

did find credible sources and the article is published on *SciJourner.org*. We have learned to never say no immediately and remain open to surprises.

There is also the challenge of the "too hot" topic. A student researching resveratrol, a substance found in red wine that is credited with extending life, uncovered what she thought were recent references (within the past three years). However, it turned out that important new research on the topic had emerged only months ago. In short, if the topic seems to be "hot" it may be worth telling students from the start to check press releases on *eurekalert.org* or *sciencedaily.com* at the beginning.

Perhaps the most intriguing and difficult of all topics is the "investigative report." These stories are challenging even to professional reporters. A story on the dangers of a local waste site or the failures of city hall to deal with lead paint needs to pass the "is it doable" test. Some investigative stories can take a year to write. On the other hand, debunking an urban myth such as alligators in the sewers is within reach of most students.

Personally Meaningful

Here's the good news: finding topics that interest young writers is really pretty easy! Science is in everything—sports, hygiene, hobbies, current events, gardens, a family's health, and much more. An avid gamer wrote a story about the connection between *World of Warcraft*, an online fantasy world, and epidemiology. A sports fan explored how technology could improve the reaction time of a soccer goalie. Numerous examples are available on *Scijourner.org*. Indeed, the most powerful question a teacher can ask when a student pitches an idea is "Why are you interested in this topic?" The story behind the pitch is often the better article.

For those students who claim to have no ideas or interests whatsoever, we play a game called "What do you do when you don't do homework?" A part-time job as a waitress can lead to discussions of food safety; swimming as a sport might trigger a story on the advantages of saltwater pools; painting might suggest a story on the advantages and dangers of cadmium sulfide yellow. One student, thinking he had us stumped, simply said when not in school he slept. "Great," we said, "There is lots of interesting sleep research." Again, helping students to understand that there is science in everything is a major step forward in promoting science literacy.

We have also seen students feed off of a teacher's passions. A story on underwater sea vents was generated by a biology teacher who took students scuba diving in the summer. Never underestimate the power of an inspired teacher!

As we have said, topics should be relevant to the author and born from a personal connection, concern, or curiosity. For example, if a teacher assigned "deep vein thrombosis," a condition that affects teen athletes and can lead to a potentially fatal embolism, few students would touch it. Yet, we have published this topic on *SciJourner.org* because the author's mother worked with patients who suffered with the problem. Moreover, the student found a classmate who

Figure 6.1

Topic Idea Board

Students repeatedly mention that topic selection is the toughest part of the writing process. A community of journalists within a classroom will offer support for students to share, consider and inspire the topic ideas of one another. Here, topic ideas are posted in a classroom as they are explored and eventually evolve into stories.

had been treated for deep vein thrombosis. It was a topic that was personal for the writer and as such she assumed a level of ownership.

Tapping into student interests can be done in several ways. Often, showing young people the stories completed by their peers at *SciJourner.org* is enough to get them to begin thinking about their own interests. Teachers can also model their own thinking about topics. Work alongside your students, staying a day or two ahead of them. Show them the topics you considered (consider at least three) and then tell them why you picked the one you did. And if it doesn't work, show them how you move on to the second topic. If you have a topic of national or international interest, show them how you might find a local angle. Show them the variety of topics that other SciJourn students and teachers have developed.

In a recent teacher workshop, for instance, one participant chose to write about the newest game technology, another about a drug company in which he was considering buying stock, another about the chemistry of the decals used on McDonald's give-away cups, while yet another focused on the importance

---

of dental care for the physically handicapped—a story which combined her inside knowledge of being a dentist's wife and having a handicapped sibling.

Of Interest to Others

As our professional editor tells students, "Your teacher is paid to read what you write. Readers of *SciJourner.org* are not." His point, of course, is that students need to find topics that will interest other teens or they have to make what interests them compelling enough so that others will want to read it.

One good way to drive that point home is to have students read *SciJourner. org* or other publications to which you have access. Talk about why certain stories interest them and why others do not. Are there certain articles that caught everyone's attention? Are there certain ones that caught no one's attention?

Figuring out what is of interest to others presents a situation where students really know more than teachers. There is one caveat: If there are bullies in a class, they can stifle conversation. In this case, divide into smaller groups or find other ways to keep the conversation civil. Again, the question "Would anyone read this article?" often backfires. Why? Because the question, as framed, requires a yes or no answer and insecure students simply reply "no" and abandon a topic idea that could work nicely. Instead, we have used our favorite question "Why are *you* interested in this?" to help students identify a good angle on a topic they really like. This question opens a dialogue with students rather than shutting them down, and the answers can be fascinating. A topic that is going nowhere suddenly comes alive when a student explains the root of his or her interest.

This Sounds Like Trouble

Student: I want to survey on drug use at our school.

Teacher: What drugs do you have in mind?

Student: Alcohol, cigarettes, and marijuana.

Teacher: A good news story, but one that will get us all in trouble. If you accidently release the names of students who use these drugs, the school could get sued. Can you think of a better approach? What about that proposed referendum on making medical marijuana legal? You might discuss the science, which would be informative.

The point here is that all ideas need to be considered in context. The teacher and students are part of that context. So is the school's administration. This one is tough; encouraging bravery is also a part of journalism. In one school where we worked, wrestlers were covering up a contagious skin infection with makeup so that the team could maintain its winning record. While one student wanted to pursue this story, he was virtually shouted down by other kids in the

room. Whoever said teaching was easy? From a journalistic perspective, this was an excellent topic idea. The decision to expose the wrestling team's bad and frankly dangerous behavior was one that needed a full-scale discussion.

Curricular Connections

Some teachers feel that the only way they can participate in SciJourn is to focus students on topics relevant to the subject they are teaching. Certain subjects—for instance, environmental science, health, or wildlife biology—are easier to work with than others. Physics, where the emphasis is on fundamental principles and mathematical relationships, may be a more difficult stretch.

To help students in hard-to-develop subjects, the teacher may want to identify "areas of interest," for instance, sports, medicine (x-rays, MRI and artificial joints all have interesting physics), building construction, or the weather. A different strategy is to simply use parts of the SciJourn process, such as the read-alouds discussed in Chapter 5, and wait until another year to get students to write the full-blown story. A third idea is to ask students to report on his or her own (or someone else's) original research. See Chapter 8 for some good ideas. If the students are doing science fair projects, this might be another good source of stories, especially if the project is put into context, as discussed in Chapter 9.

A good way to practice curricular reinforcement is to begin by asking students about some of the science topics they studied during earlier years in school. Are there any topics they wish to learn more about? Teacher Rose Davidson, wanting to stay close to her chemistry curriculum, had students randomly choose elements to build a story around. "Some students felt they had very easy and interesting elements: fluorine in toothpaste, chlorine in swimming pools, and helium gas in balloons. Other students had to stretch to find context: rubidium in fireworks, argon in lightbulbs and lasers, and carbon's role in global warming. By the end of the class period most students were on the trail to their topic."

Four Pitches: Warm-Up, Basic, Power, Speed

In newsrooms, pitching presents an opportunity for the author and editor to discuss the angle of the story, to narrow the focus of the topic, and to explore the feasibility of completing the story before the deadline. During the process the editor serves as a surrogate reader, probing the journalist with questions and offering possibilities about ways the particular story idea might be covered.

We recommend you institute something similar in the classroom. It is during the pitch that the teacher can celebrate good topics, weed out bad topics, help focus ideas and get the student going.

Strategizing the Pitch

We have identified two basic modes of pitching, one we call the *warm-up pitch* and the other the *basic pitch*. Two additional strategies, the *power pitch* and the *speed pitch*, can be done after students have conducted some preliminary research.

The warm-up pitch is a riff on a strategy teachers know as "the fishbowl." Two chairs are placed facing each other in the center of the room with remaining chairs or desks circling around them. To get the classroom pitch going, the editor (initially the teacher) sits in one of the two chairs, and the student journalist sits in the other. The conversation will vary, depending on whether the student journalist is someone with a quick, ready-to-go idea, someone with a hidden good idea, or appears to be someone with no idea whatsoever.

Typical editor questions include

- What is the topic for your story?

- What is the angle? How will the story be made interesting for the readers? Why are you interested in this?

- How will you incorporate science?

- Why does this story matter now?

- What sources will you consult? Are they reliable? Will there be any real experts to talk to? Can you do any original reporting?

Following this conversation, the journalist changes seats and becomes the editor and a new student comes forward to pitch his or her topic. (A scripted version of the warm-up pitch is available at *www.Teach4SciJourn.org* for those who need more details.)

The warm-up pitch takes a lot of time. If you are working with a small group, it is viable, but for the normal-size science or English class, it needs to be modified. After hours spent in classrooms listening to warm-up pitches, we developed what we now call the *basic pitch*. While seated in a circle, each student briefly describes his or her topic to the class. The teacher moderates and fellow students probe for more detail using questions such as those listed previously.

Typically, SciJourn teachers have started using the warm-up pitch for several students and then moved on to the basic pitch for the remainder of the class. Whether you go for the warm-up pitch, the basic pitch, or some combination thereof, it is important that students are prepared to modify or even change topics as they delve into their research. A story that began with post-traumatic stress disorder evolved into how the drug propranolol was helping reduce stress in many other situations. Creating a low-risk environment is critical to encouraging student reflection and a willingness to revise topics.

Some teachers may prefer what has come to be known as the *power pitch*. The students select a topic and do some preliminary research, which is then compiled onto a few PowerPoint slides. Each student then presents the slide presentation to the class for feedback, inviting many of the same questions and responses as in other pitch forms.

The SciJourn teachers who developed this approach—Linda Gaither and Patricia Baker—critique the PowerPoint slides with questions such as

- Says who?

- What sources?

- Where are you going with this?

- How is this interesting to teenagers?

- Do you know anyone who has this disease?

- Really, that is *all* you found in a whole class period?

An advantage of the power pitch is that the student has already done some legwork on the topic (and may have already abandoned topics that are unworkable) and has at least a few details and sources about which classmates can comment. We have even seen classes where teachers allow students to swap topics with one another after power pitching. Power pitching also works to move students along in the writing process (see Chapter 10: Going the Write Way).

The *speed pitch* is modeled after speed dating. Students work in teams for a set amount of time. It begins with students sitting in pairs, opposite one another. The outside person pitches their topic as the inside person acts as editor. The outside person doing the pitch writes down the suggestions and ideas. At the end of a few minutes, the inside person pitches while the outside person becomes the editor.

For round two, the inside people move one space clockwise and the process is repeated. Three rounds are completed and then each student uses the remaining time to work on the other spaces on their speed pitching paper, the one where they jotted down the ideas and questions they collected. As a check on the process, the teacher can collect the papers at the end of the hour.

Regardless of style, pitching is useful for another reason. As the class becomes familiar with one another's topic there is the potential for sharing of websites, photographs, names of experts to be interviewed, and so on. Moreover, it helps establish a community of practice. On a real newspaper or magazine, reporters will often "bounce ideas off of their editor" as they pull a story together. The key is that the advice must keep the story focused.

Responding to Pitches: Dos and Don'ts

- *Do* recognize a great idea and leave it be. Lots of students will just get a great idea right away, and it's important to congratulate them and let them move on to the research phase. Sometimes, these great

ideas come from off-beat or low-achieving students and this, at times, has frustrated the high achievers. A kid who was tattooing himself wanted to do a story on the health risks of tattoos, another who had cochlear implants used *SciJourner* as an opportunity to explain how the things in his ears really worked.

- *Don't* suggest broadening a topic. The single most common mistake teachers make in "helping" students is taking a narrow topic and suggesting that students enlarge it. Textbooks are about broad. Curriculum is about broad. SciJourn is about the specific, the concrete, the personally relevant and interesting. Only by staying narrow and focused can students really dig into the science of their topic. Listen for the specific and when you hear it, congratulate the one making the pitch and then tell other students why this is a good topic—because it is narrow enough to research and write about intelligently in a short article.

- *Do* look for the story behind the story. Some students come up with topics that appear to be too broad but there is really a better story behind it. A student planning to write about cancer was really interested in his grandmother's renal cancer brought on by years of smoking. That is the story we told him to write. Remember, asking the student why they are interested in the topic can often elicit these more meaningful stories—what journalists call an "angle"—on some bigger topic.

- *Don't* push your ideas for a story or immediately reject a student's pitch. Engagement is key to success. Rarely does the teacher's idea for a story generate the same level of energy as one the student dreams up.

- *Do* respect your students' ideas. Sometimes, that really horrible, unbelievably stupid idea for a story turns out to be a surprise. For example, a pitch on how titanium-containing necklaces increase performance in sports generated a heated argument between the chemistry teacher and a student. The teacher adamantly argued that the necklaces were junk and the student, just as adamantly, argued in favor of their powers. "Show me that I'm wrong," said the teacher. Sure enough, a few days later with some research in hand the student came back saying the necklace was indeed a phony and now wanted to write an expose!

- *Don't* settle for "old" stories if you can help it. For example, Pluto's demotion as a planet still riles kids, but the issue seems settled and astronomers' attention is now focused on finding planets outside our solar system. Push your students to see science as a dynamic field where change is valued.

- *Do* make the stories timely. One of the first *SciJourner* stories we published was about the difference between a heart attack and cardiac arrest, which was news following the death of entertainer Michael Jackson. Earthquake stories are not up-to-date until there is an earthquake somewhere. Journalists know how to make "old" stories new by tying it to a recent event. That Pluto story could be new again if astronomers spot an oddball "planet" that doesn't fit their standard description. Asthma becomes new again during National Asthma Week or when someone in the area is suffering during soccer practice. Journalists call these refreshable topics "evergreens."

What Does It Look Like?

To facilitate the teacher's/editor's role in the basic pitch, we present these 10 classic dialogues designed to help you offer thoughtful and quick responses. The trick is to identify the issue at hand and make your move.

Go for It!

Student: I've noticed that the number of feral pigs in some of the state's southern counties has increased. My uncle, who owns a farm, says that they are becoming a big nuisance because they tear up the land rooting around for food.

Teacher: Wow. I've never heard about this. That's a great story. I hope you'll talk to your uncle about it. And can you find anyone else to interview, such as a local farm agent or wildlife biologist?

Student: My uncle knows the local farm extension agent. I could talk to him.

Teacher: Great. This story is really coming together! Go for it; let me know if I can help. Be sure to check out credible sites for information on feral pig biology and whether they are a problem in other places.

When the Topic Is Too Big

Student: I want to write about cancer and its causes.

Teacher: That is an awfully big topic. Why are you interested in it?

Student: My aunt has lung cancer. She smoked for 20 years.

Teacher: I'm sorry to hear that. However, if you could interview her, it would be a better story to talk about her lung cancer. Can you talk to her?

Student: Yes.

Teacher: Then you make the story about her by starting with her story—how she found about the cancer, how it affects her, and how she is being treated. Then you go into the medical science—what is lung cancer, the symptoms, treatments, outlook, how many Americans are diagnosed with it each year, and so on. You end the story with your aunt—how she is doing now and coping. So it's the human story that interests the readers and "sandwich" the science in between. Sounds doable?

Student: Yes.

Teacher: Are you ready to do this story?

Student: Yes.

Teacher: It will be tough, but it is often therapeutic for someone with a serious disease to talk about it. In a sense, you are doing a good deed. And it is a really interesting topic.

When the Topic Is Too Vague

Student: I'm really interested in astronomy, but I'm not sure what to write.

Teacher: Well, that is a big, but interesting topic. We need to find something narrower. Anything in particular about astronomy?

Student: I think the Hubble Space Telescope is pretty cool.

Teacher: Yep, but that is still too big a topic. You could write about an instrument on the telescope, especially one that has been upgraded recently during a Shuttle mission. Or a finding made by the telescope. If you search Hubble Space Telescope news, you may find something recent and interesting. Or you can go to *eurekalert. org* and find a recent press release.

Student: OK.

Teacher: Anything you know now that is interesting to you?

Student: I read that the telescope has found stars being "born." I would like to investigate that.

Teacher: Let me know what you find. You don't have to discuss all the places they found new stars; one or two will do. This will not be an easy topic. You are going to have to do some serious research, but it is indeed an interesting topic. We might even find an astronomer at the science museum to talk to. We can brainstorm on other ideas if this doesn't work out.

Is the Idea Credible?

Student: I want to write about how color affects your mood.

Teacher: Hmm. That may be a hard topic to find credible sources. Have you found anything on the web?

Student: There is a website that lists which color goes with what mood.

Teacher: Who is behind this website? Anyone credible?

Student: I don't know.

Teacher: Read the "About Us" on the site and see who the author is, whether it is reviewed by experts and is regularly updated. All credibility clues. If there is no About Us or it is selling "mood enhancers," then the site may not be credible.

Student: OK.

Teacher: And I have to warn you that this can be a topic with little real science. Be critical and be willing to abandon it if you can't find credible sources.

The Student Appears Clueless

Student: I have no idea what to write about.

Teacher: Do you have any hobbies or outside interests? What do you do when you don't do homework?

Student: I like to wrestle. I'm on the wrestling team.

Teacher: Great. There is a ton of science in sports. Anything about wrestling that you want to investigate? Are wrestlers at greater risk of some disease or injury? Any long-term effects from wrestling? Is there a skill that wrestlers need to master that might be worth understanding in more detail?

Student: Because wrestling is a contact sport, we get skin diseases, like ringworm or a form of herpes.

Teacher: Good topic. You could interview the coach or trainer, talk about how the disease spreads, what it looks like, symptoms, and how to prevent it.

Student: Yeah. We can't wrestle if we have any sign of a skin infection. The coaches and referees keep an eye out for it.

A Quick Method for Telling Who Feels Ready With a Topic … and Who Doesn't

Tell students to think of the front of the room as a line measured off in gradients 1 through 10. On the 1 end (get up and demonstrate) are people who have no idea whatsoever what they want to write about. Clueless. Blank. On the 10 side (move across to the other end) are people who are absolutely clear and do not feel that they need any help at this point. They are ready to research. Then ask students to line up where they feel they belong.

Within seconds the teacher can get a sense of who needs help and who doesn't. If you need a moment to jot down names, ask students to chat with the people nearby to see if they in fact should be moving a bit in one direction or another.

This activity can be repeated as often as need be. And you need not limit yourself to one topic. Have people line up in terms of who feels like their drafted article will be done by tomorrow, for example.

Teacher: Seems like a good topic, and don't forget to check out some of the credible medical websites, such as WebMD or *mayo-clinic.com*. And check out the amateur wrestling association websites; they may have information as well. All credible sites. You could even talk to a wrestler on the team who has had a skin disease; that would be interesting. I think this could be a good story.

Clueless 2

Student: I don't know what to write about.

Teacher: What about a hobby or outside interest? What do you do when you aren't doing homework?

Student: I work.

Teacher: Where?

Student: I'm a waitress at the Lotus Garden, a restaurant not far from here.

Teacher: There is a lot of science in handling and preparing food. Any particular issues you deal with—problems or challenges keeping the place clean, any particular food issues, something that has come up lately?

Student: Well, the recent outbreak of flu has gotten my manager all concerned. She has us wiping down the menus, tables, and chairs with a special cleaning solution.

Teacher: Sounds like science to me. It would be interesting to see what is in the cleaning solution and think about restaurants as a source—doctors say vector—for spreading disease. Do you have any health department regulations?

Student: I don't know, but I could ask my manager.

Teacher: Great. That could be one interview. You could also contact the county health office and see what they say. Have they put out a special warning? Also, check out the medical websites before you talk to the county. You need to find some background information on this particular flu—how does it spread, how dangerous is it? The CDC tracks flu outbreaks and their website is full of highly credible information. This should be an interesting story.

Clueless 3

Student: I don't know what to write about.

Teacher: What do you do when you aren't doing homework—any

hobbies or sports?

Student: No. I just work on cars with my father.

Teacher: Well, there is a lot of interesting science and technology in cars. What do you do with cars?

Student: We fix up old cars and sell them. I'm saving up to buy a car of my own and pay for insurance.

Teacher: Good for you. Say, has car technology changed a lot over the years?

Student: Yeah.

Teacher: You might explore that. How easy is it to tune your own car these days versus 30 years ago? Something like that. It is sort of a good news/bad news situation. The new technology is really sophisticated and requires special equipment, but cars are cleaner running and more fuel efficient. Or you could focus on a car system—how can you improve fuel efficiency and the trade-offs. Is there anyone you can talk to?

Student: I don't know.

Teacher: What about your father?

Student: He only has a high school degree…

Teacher: …but probably 20 years of experience fixing cars. In this case he is more an expert than any MD. Expertise doesn't always mean an advanced degree. The other information can be found on the web. I'm sure that there are websites that talk about horsepower, performance, and fuel efficiency. Sound good?

Student: Yeah, I think it does.

When the Story Is Good, But the Direction Is Not

Student: I get terrible headaches and my grandmother recommended that I try "cupping" to help. It works well. So, I want to write about a story on why teens should try alternative medicines.

Teacher: Interesting topic. Did you first see a doctor and try standard medicines?

Student: Yes. For years and it didn't help at all. Only cupping makes a difference.

Teacher: OK. That is interesting but I need to warn you that you need to back up your statements with credible sources. There is a National Institute for Alternative Medicines—be sure to check them.

Student: OK. I just want folks to understand how much this has helped me.

Teacher: That's great. However, you need to be honest. There are alternative approaches that don't work. You need to give a balanced accounting. We don't want to say all alternative medicines are great; not all prescribed medicines are great or even work for everyone, as you found out. And if cupping is not well studied by scientists, then you need to say that, too—this is an experimental approach.

When the Teacher Doesn't Understand the Hook

Student: I am going to write an article about Polaroid cameras.

Teacher: Why do Polaroids interest you? That's a pretty old technology.

Student: Lady GaGa is doing a lot of cool stuff with Polaroids.

Teacher: Great lede. Go for it.

The Double-Duty Double Cross

Student: I want to write about oil paints.

Teacher: That sounds interesting, but it is a pretty big subject. Why are you interested in that topic?

Student: Last year, I wrote a paper for my art class on oil paints. It got me interested in the topic.

Teacher: Was it a science paper?

Student: No. It was more about what colors are available and the different ways to apply paint.

Teacher: Well, we need a more science-based question. Is there anything that came up that might suggest a science article?

Student: My teacher said that artists were being told not to use certain paints because they are dangerous. They cause cancer or make you ill.

Teacher: Good. I don't think you need to cover all the "bad" paints, but you could pick one example and discuss its dangers. You could also see if there are safe replacements, and if they work as well as the original paint. You could even interview your art teacher.

Student: That seems interesting.

The Editorial

Student: I want to write about how everyone should buy only hybrid cars.

Teacher: While I can imagine that there would be some good science in that story, it does seem like you are writing an editorial. Do you know the difference between a news story and an editorial?

Student: Not really.

Teacher: A news story reports what experts who know the technology and people who use the product think about it, while an editorial takes a position on some issue of importance, like arguing that we should raise the age for driver's licenses to 21. In this project, I want you to be a reporter.

Student: How do I do that?

Teacher: For this story, you should weigh the benefits and disadvantages of hybrid cars versus gasoline-powered cars. For example, hybrids are more expensive to buy and need new batteries every few years, but they use less gasoline in city driving. Maybe, a small car that gets 30 mpg in the city could be a better buy and also environmentally friendly. Or if you work on a farm and have to haul hay and feed all day, a truck may make more sense. As a reporter, you get the facts and opinions, but you never take a particular side in the discussion. That is what I want you to do.

Other Ways to Find Topics

When on a tight time schedule, covering local events and touring labs has been an effective strategy. Something is usually happening at the science museum, zoo, aquarium, or other local institution with a science bent. Teens can register as reporters and cover the event, just like a local TV crew might. There is no official organization that licenses journalists. If you act like a journalist, you are a reporter. That means that if your students are ready, they can attend press events or tour facilities as reporters, too. An hour or so of work typically yields enough material for a group to pull together a story within a few hours.

The same approach can work if you wish to cover the local science fair or the neighborhood health clinic. Because the SciJourn project is near an urban area, our lab visits target universities, medical schools, nonprofit research organizations and companies eager to recruit potential science students. Typically, they arrange a wonderful tour, showing off their coolest devices and experimental setups.

Nearly every organization of size has a public information officer (PIO) whose job it is to help the press cover their institution. Typically, the PIO can be

---

found on the organizations' website—search for "press"—or try a cold call to the organization. Universities, museums, public utilities, and medical schools may even have outreach programs that work with local schools. It is our experience that almost all organizations love the idea of teen reporters, even asking for our stories for their own public relations efforts.

Obviously, it is important to prepare ahead of time for an event or tour. If there are speakers or presentations, we typically assign a team to cover it. One may be the reporter, asking questions (see Chapter 8 for interviewing tips), while another takes photos or video. Make sure that the electronics are fully charged, extra batteries are available, and reporter notebooks and extra pens are on hand. Despite how many times we do this, someone forgets something. Chapter 12: Beyond Words explores video, podcasts, and photography.

Whether it is an event or a lab visit, it is important that your students identify themselves as reporters. We all have an expectation of privacy, but talking to a reporter negates that unless the person says no. It is all on the record, which means that anything they say can be in the story. The ID badges you see reporters wear are nothing special, just something that says reporter. You can make your own, which is what we have done, or just tell students to declare themselves as reporters.

After the event, it helps to thank your contact and make sure to send anything published their way. Don't be surprised if you get an invite for next year.

CHAPTER 7

FINDING AND KEEPING TRACK OF SOURCES

In elementary or even middle school we work hard to help students understand that science is not what is in the book, and that science ideas emerge from systematic observations and experiments. We stress hands-on/minds-on learning, not simply because activity is more memorable than descriptive language, but because predicting and generating ideas from repeated trials is what science is all about. And then we move into high school or become adults. Suddenly, the science we wish to understand is most often not available to us through hands-on investigations. Ideally, we still understand that the information provided to us in books, articles, or on the web is based on experiments and systematic observations, but we are no longer making most decisions regarding the credibility and viability of ideas using our own senses and our own experiments. As individuals we often need to make judgments based on secondhand descriptions, text, or conversation.

Science journalism promotes science literacy precisely because it demands that readers and writers thoughtfully consider which ideas seem most convincing given the information at hand. The information at hand comes from what journalists call "sources."

The word *source* actually refers to two things: the place where information originates and the place where information is found. The place where the information is found may or may not be close to the original source of the scientific undertaking. In schools we tend to teach sources as synonymous with citation. "How many sources do we need, Mr. Roberts?" The citation is actually just a directional arrow that tells the reader where the information included in a text comes from. In this book and in the journalistic community, we use the word *attribution* to refer to this directional arrow, the crumbs one can use to follow information back to the place the author used as a source. Again, the source is the place; the attribution is the directional arrow.

In the landscape of science, information flows from the researcher's lab or site of field study to various quarters. Typically, the scientist writes up his or her findings for a peer reviewed journal. In for-profit entities, it is often considered smart not to share research until it is patented. Thus, most science is only read by a small number of experts.

However, some science information is deemed newsworthy. In most universities, government agencies, and companies, the big finding or commercial product is funneled through the press office. Press offices work to highlight the research or item and shine a positive light on that institution, agency or company. Even research funders, such as NSF or Howard Hughes, get into the act and generate their own press releases.

Press offices understand what it takes to get research reports into the media. One news magazine targets readers who are scientifically informed and the press office shares information that it deems scientifically important, while another, more concerned with popular appeal, is pitched stories that grab attention. One source on the web exists primarily to sell "stuff," while another is funded by a foundation to share the latest information on a disease. All share some scientific information, but what is important, how it is presented, and the material's credibility varies.

In the Classroom

In this chapter on web searching and the one that follows on interviewing, we discuss the credibility of sources and various methods people have used to help students differentiate among and evaluate sources. Our goal in both chapters is to encourage critical rather than mechanistic thinking.

Scientifically literate people seek credible sources, knowing that they cannot always depend on firsthand encounters. Several credible sources are typically needed to assess the value of new findings or commercial product in order to confirm information that is under scrutiny. Examining several sources also

helps identify possible disagreements or ideas still being challenged. It is not enough to simply assert credibility—"I read it in a book" or "Scientists have proven that...." Instead, a scientifically literate speaker or author needs to recognize the difference between embryonic and consensus ideas and attribute information to specific sources. A scientifically literate individual should be ready to explain why she considers a source credible. Only then can listeners/readers determine if they agree with the author's assessment of those sources or check out the speaker's interpretation of the information that was cited.

A scientifically literate decision, argument, or statement is based on multiple, credible, attributed sources. In fact, multiple, credible, attributed sources has become, in some sense, the SciJourn mantra. For teachers who grade student papers, we know that it would be easier to separate out the three parts—give one point for multiple, another for credible, and a third for attributed. We tried that, and found over and over again that it does not work. If the sources are credible you wouldn't know it unless they were attributed. And one source, even if attributed and credible, is not enough. That's why our third standard (see Chapter 4) refers to multiple, credible, attributed sources all rolled into one. These qualities are interrelated.

Cindy, one of the SciJourn staff with expertise in searching and credibility, was invited into a school. She began by challenging students, "What would you say if I told you school was canceled tomorrow?" Not surprisingly, several students raised their hands and said things like, "Yeah, right!" and "No way would I believe you." When asked the all-important question, "Why not," the students responded with well-reasoned answers.

> Student 1: "Who are you? You don't work for the school. You could just be lying … or crazy."

> Student 2: "There was no reason that school would be canceled; there was no snow. You didn't tell us about some emergency."

> Student 3: "That just doesn't make sense. Too good to be true."

When the students were asked, "Who would you believe if they said that school was canceled tomorrow?" they universally agreed the only person would be the principal.

"Not your teacher? You know her, she works for the school, and you trust her, right?"

The students thought for a moment and one responded, "Yeah, but she's not in charge to make those decisions, so we trust her, but she doesn't necessarily know, you know?"

The students in this class had made a complex credibility assessment. Cindy hadn't met their criteria for trustworthiness. She was an unknown source, lacking credentials, with unknown motives. Her information didn't pass the common sense test based on their prior knowledge of school cancellations. Finally, she held no authority. Even the students' teacher fell short on authority, despite the students' trust. They felt that she wouldn't necessarily be the one to "know."

Teenagers make credibility decisions every day. The teacher's goal is to help them connect, and thereby value, their experience as critics of "real life" information and then connect those everyday life skills to their internet searching and credibility assessment. Their experience provides a solid foundation on which to build skills for internet literacy.

Cindy's story provides a lovely example to shape our discussion of credibility and internet searching. You might even want to use it to begin your own discussion on web searching, making sure to name and record the categories students use to assess sources of information. Point out that the students who Cindy visited know that she a credible source, but her statement that school is cancelled is not believable. On the other hand, if she told them about a new site for mounting photographs online, and they queried her about how she knew that that was an excellent site, they would in all likelihood find her convincing.

Credibility has to do with the coming together of background knowledge, purpose, and an ability to read contextual clues. Alas, the answer to the question "Which sites are most credible?" is not perfectly plain. Credibility cannot be determined by the name of the URL. But students need to understand that all sources are not equal. Journalists and other scientifically literate individuals search for the best sources for their particular query, given that moment in history and their own limited understanding of what is being said.

The bulk of this chapter focuses on web searching. However, it is important to understand that one of the journalists' main tools is the telephone. Well, not just the telephone—they interview people in person, online, wherever they can find a good source. How do science journalists locate good contacts? Who is, in fact, a good source? What do they say once they have found someone to talk to? Chapter 8 describes in detail how to do an interview. But the truth is, before students talk to anyone, they should first do background research. This chapter is designed to get them started by searching the web.

Carousel Activity

This is an activity to help students ask good questions, think about sources of information, and conduct effective searches. It works well after students have selected their topics but before they have done research. Each student gets a large piece of paper or poster board and writes his or her topic across the top. The paper is then divided into three columns with the following headings: Possible credible sources of information; What questions would you ask these sources?; How would you find each source? In small groups, students move around the room, discussing the topics and filling in the charts. At the end of the activity, students have their own charts filled with ideas to use as they begin searching.

Assessing Credibility on the Web

Students have typically been exposed to two strategies for searching the web. One sets up a list of acceptable and banned sites, while the other presents guideposts (usually built around a mnemonic) designed to help the user rate a site's credibility.

In most communities, schools are required to block certain websites either by law or through funding restrictions. While the "banned site" approach may help deal with the usual problems of pornography, music downloads, and time-killing social media, this approach creates a new set of issues. Is it really wise to outlaw a site on the benefits of breast milk because it contains the word "breast?" What about no dot coms? WebMD or Mayo Clinic's sites, both highly credible sources, are in fact, dot coms.

But perhaps more importantly, if part of the function of school is to prepare students for life, the inauthentic closing down of generally accessible information does not teach students how to function sensibly in a democratic society. We are talking here about high school students who get to drive, join the military, and often make choices that have long-term life consequences. Let's talk to them about which sites make sense and which don't.

To support the guideposts approach, educators have generated rules—many identified with mnemonics such as AABCC (Accuracy, Authority, Bias, Coverage, Currency) or SCARAB (Substance, Currency, Authority, Relevance, Accuracy, Bias)—to determine the credibility of sources. Schools have sought to block inappropriate sites in the hope of making only credible sites available to learners. Teachers also institute rules, such as no Wikipedia, for this same purpose. These are all mechanistic attempts to assess credibility. Instead of tying credibility to the source itself, our efforts have been directed toward teaching students to think of sources in context.

Setting up guideposts does seem a more sensible approach than blanket bans, especially since it involves working with students to develop criteria that can be used to evaluate websites. Who is producing this site? Is the site updated regularly? Is it reviewed? Does it seem biased or is it selling a product?

However, this critical-thinking approach is actually a secondary method, something to be used when a student stumbles into a site, like coming across an interestingly titled book in the library. The more challenging question for web searchers is where are the tried-and-true reference books? That's where you begin.

Before you send your students blindly to the internet, try modeling what a good search looks like, thinking aloud as you make your moves. An internet search-aloud/think-aloud can be done as a live demonstration, but we strongly suggest some preplanning once you have read this chapter. Becoming familiar with stumbling blocks like dead websites can save time and provide your audience with the information that describes why you are approaching the problem in the way you are. A lesson plan detailing one way to conduct a search-aloud/think-aloud is available online at *Teach4scijourn.org*.

Background Knowledge

Reporters, unlike most of your students, have sophisticated knowledge about how the world of information works. They know, for instance, that many of our national medical statistics and much of the information on health comes from the Centers for Disease Control and Prevention (CDC), the National Institutes of Health, and major research institutions and hospitals. When information shows up on *ehow.com* or *about.com*, they recognize that this is *a* source, not *the* source. The numbers that appear on these commercial sites are typically taken from the more credible sources. Again, the information is not necessarily wrong, but these sites often do not cite the original or true source that can be checked or queried. Nor do they have a commitment to be up-to-date.

Students have much more limited experience with the world. When asked, high school students in one classroom told us that Columbia University is in the country of Colombia, that a graduate student is someone who had graduated from grade 12, and that Mountain Dew works as a contraceptive "if you drink enough of it." The point here is that as educators we have an important obligation to unearth similar misconceptions and expand the basic knowledge—not just the science knowledge—of young people. This is one of the places where students whose dinnertime conversation includes discussion of colleges, and whose family vacations and other nonschool activities introduce them to background information that supports their school learning are at a distinct advantage.

The bottom line is that to level the playing field as much as possible, you may need to simply tell students that journalists and other scientifically literate people have information in their back pockets and that it is part of your job as a teacher to help fill *their* pockets. Encourage them to ask questions and be generous with your answers. Query them when you suspect a misunderstanding. Push them to look for the highly regarded sources rather than whatever comes up first on the search engine.

Students' initial tendency is to be happy with almost any source ("I got one. Let's reel it in!"), while journalists seek *the* source. They are trained to ask, "Where does this information originate?" If you needed the population of Massachusetts, where would you go? Google it and a number pops up on lots of sites. However, the agency charged with determining the number (and the credible source by newspaper standards) is the U.S. Census Bureau, located within the Department of Commerce. They are the folks, who by constitutional requirement, are given the funds and have the mission to generate this number to the best of their abilities. That doesn't mean the number on other sites is wrong; it just isn't the real source. Just as forensic accountants learn to "follow the money," scientifically able people need to learn to follow the data back to its source. In fact, the best data on a new battery design may well be on a dot com.

Students may do well to think of this like the children's game "telephone" or the way rumors spread in a school. If you heard a rumor about yourself, would you stop at the first person you heard it from? Or would you continue

Figure 7.1

Search Terms: Experts Versus Students

Cathy Farrar, a researcher on the SciJourn team, asked respondents to think about search terms and websites they might use to better understand two topics—diabetes and volcanoes. "Wordles," a visual representation of the search terms and their frequency, show the difference between the terms and strategies employed by (a) a group of scientifically literate adults and (b) a group of high school students before they worked with SciJourn. Farrar's research studies on scientific literacy are cited on *Teach4SciJourn.org*.

(a)

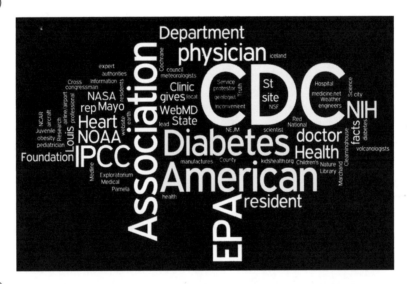

(b)

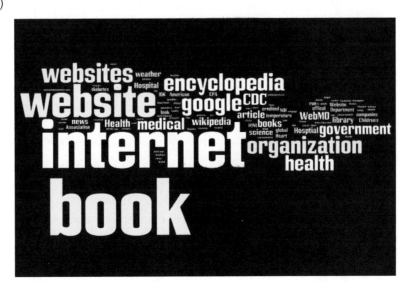

"I remember being shocked that my students did not know what the Mayo Clinic was or what the letters CDC stood for. They didn't even know what the CDC did and they had not even heard of the FDA. So to them these sources were not credible sites. I understood that in their world these organizations were not a part of their everyday language. So I explained to them that I had worked for the FDA and I had them visit the websites of the Mayo Clinic, CDC, and FDA. We had multiple conversations about what these organizations do and why we use them as creditable sources." — Tonya Barnes, SciJourn chemistry teacher

to follow the trail of information until you found the original offender? This is also a chance to teach a fundamental lesson about the web. It's all about links and trails, hence the name World Wide Web. Disentangling the web, sorting through tiers of offshoots and tangents to the original authoritative source is key.

In the course of implementing SciJourn, we have seen teen authors attribute information without worrying much about the credibility of the source, even when we prime them to use American Cancer Society or CDC. This is the problem with starting with Google or another search engine as the first step. Sometimes students think that what pops up first is the site of choice and as far as they will ever need to go.

Searches and Sources

Our approach is summarized in Figure 7.2. Here is a more detailed rationale:

1. Give the students a beginning list of sites you view as highly credible and useful for checking out science information. Make these lists specific. If you know that one of your students, for instance, will be writing about wind turbines, list sites like *EPA.gov* (Environmental Protection Agency), *NYTimes.com* (New York Times), *ScienceNews.org*, and *awea.org* (American Wind Energy Association). Then invite students to add to the list as they encounter sites they consider highly reputable. Discuss their nominees for additions to the list. You may consider creating a version of this list during read-alouds as you encounter and discuss sources included in published articles.

2. Carefully explain (repeatedly) how information is generated, what we call the "landscape" of science information. Knowing where information comes from and how it got there is vital. The internet is filled with "content farms"—sites that post a plethora of information on a vast array of topics in order to collect money from targeted ads that surround the content. The business of these sites is selling ads; they also know how to be the first sites listed on a Google search. In addition,

Figure 7.2

Student Guide to Finding Useful Websites

1. Check your teacher's list of generally highly credible websites to see if any will work for you. If not, generate a list of your own before trying a Google search.

2. Use Wikipedia to find specific, scientific search terms; also use their references and external links.

3. Use the most technical terms available for a general web search. Choose websites that are the source of information, not just places where information is repeated. Follow the information trail. Ask yourself, did the people behind this website determine this information or are they just reporting it? You may need to ask someone for help.

4. Try some of these Google tricks: mix your topic search terms with "organization" or "news" to find targeted results, using the Search Star shown in Figure 7.3 on page 97.

5. When the search uncovers a website that is not obviously "highly credible," ask the following questions:

   a. Does this website contain an "About us" or "About this site"? If so, read it for clues to the site's purpose and credibility. No clues as to who wrote the information? Then skip this site.

   b. Is this website "peer reviewed" or "editorially reviewed"? Is there a mention of "reviewed by" or something similar? Are the reviewers qualified? Is there a board of directors listed?

   c. Does this website reference highly qualified sources?

   d. What is the purpose of the website—to inform, sell a product, persuade, mislead or entertain? A dot com can provide good information, but it won't tell you what is wrong with its product. Similarly, dot orgs are great sources of information, but if they are advocating a cause they are not going to articulate the "other side."

   e. Is the website updated frequently? Is there a mention of "last update"? Is it recent? Science, medicine, technology, and statistical information can change quickly.

   f. Double-check the information on other sites. How does it compare? Even credible sites can get it wrong or post old statistics.

   g. Never trust sites that promote mystery "experts" or "institutions" or seem tied to a single person (Dr. Bob's miracle diet); have "dead" links or all links go to the same website; disguise product advertisements as news; claim conspiracy theories that hide the truth; or cite only unpublished results or personal stories.

---

they know what questions are frequently asked and how naive search-ers tend to ask for information. They are very good at answering obvi-ous questions in clear, definitive, easy-to-understand English. Even if the information is correct at this moment, will it be updated? Will it be a good place to look 15 years from now? What are the site owner's incen-tives for reviewing and updating their material? Some sites provide a mechanism for feedback, but who is giving feedback?

On the other hand, *nih.gov* (National Institutes of Health), *allaboutbirds. org* (Cornell's Lab of Ornithology), *umm.edu* (University of Maryland Medical Center), *noaa.gov* (National Oceanic and Atmospheric Asso-ciation), *nasa.gov*, *cancer.org* (American Cancer Society), *diabetes.org* (American Diabetes Association) and many of the other sites backed by well-known organizations, research institutions, or the government are reviewed by boards of experts. It is their business to stay up-to-date. Because most sell nothing and some, such as NASA or CDC, depend on the public's goodwill for their existence, credibility is their chief asset and what they really have to promote. These sites have a strong incen-tive to make sure that their information is accurate and timely.

3. Begin the web search by thinking of the target. After approving the author's topic, the first question should be: "Where are you going to look for information?" We push students to think about credible organizations, often as part of their pitch conversation (see Chapter 6) and certainly before they call up search engines.

   For example, an environmental topic generates the question: "Who is responsible for environmental protection in the United States?" and directing the student to *EPA.gov*. Is this agency credible? A follow-up question might be to ask what other environmental organizations students know, leading them to sites such as Greenpeace, Sierra Club, or the Environmental Defense Fund. Could these organizations have certain biases? Who would represent the "other side"? That query might lead to a search for industry groups.

4. Do a "smart search." Sometimes, the topic does not lead to such an obvious target as *EPA.gov* (for instance, the question, "Are there different types of tennis balls?"). Then an information aggregator may be a first good step. In this case, we strongly recommend a Wikipedia search, but other sites may function just as well. Typically, we tell students: "Wiki-pedia (or *ehow.com* or *about.com* or…) is a good place to start, but a lousy place to end." In this case, Wikipedia introduces us to the International Tennis Federation and the U.S. Tennis Association, which leads the student to a more directed search. According to the International Tennis Federation, there are three types of tennis balls used in professional play. The site then provides the first clues as to why they differ.

The key to a smart search is also finding the right terms. A student in one class wanted to know why "some poops float and some sink"—a gross but certainly teen-relevant topic. However, searching "poop" or its various colloquial variations, as you might imagine, does not lead to the most credible or savory sites. Information sites, such as Wikipedia, or other sites seeking some level of dignity are good places to discover the technical terms that can aid in searching. In this case, poop leads to *feces* and *stool*, the terms that gain entry into medical sites.

5. What happens when student reporters encounter sites that don't scream credible or even look a bit suspect? For example, one reporter cited *potterssyndrome.org* in her search for information on the disease Potter's Syndrome. The site is decorated with teddy bears on its homepage. In fact, the site is run by a mother who lost a child to the disease. By reading the "About Us," looking at where the site gets information, and seeing if there are regular updates; we determined that this is a credible site.

6. The Search Star (Figure 7.3) highlights another means of guiding students in their quest for information. It can be considered a more sophisticated way of conducting a Google search. Coupling a topic with specific key words will direct students to sites that will be particularly useful and perhaps difficult to find by other means. For example, to research topics involving mold, a student might Google

Figure 7.3

The Search Star

The Search Star is one strategy for more effective internet searching, but the resulting sites still need to be checked for credibility.

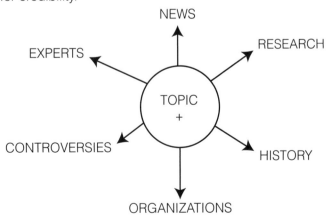

"mold news" to find current information, "mold research" for the latest studies, "mold history" to contextualize the topic, and "mold organizations" or "mold experts" for more practical information and perhaps local sources of information. The websites still need to be filtered to separate credible sites from the doubtful, but the approach can help find more targeted sites. The Search Star quickly gives students those search words. An important caveat: If students use all of these words, they could spend much of their lives searching and never writing. Avoid the formula. The star is just there to help when you have lost your way. Keep your eyes on the path, not on the star!

If You Lose It, You Can't Use It

"Learning the hard way" is a proverbial prescription for meaningful and memorable experience, but when frustration and inefficiency affect success, it's time for teacher and technological help. We have seen numerous examples of students who, when asked to include missing attributions in their articles submitted for editing, respond that they don't remember what those sources were or where they found them. To complete their articles, these students have had to go back and begin their research anew in order to retrace their steps. It is frustrating, inefficient, and a terrible waste of time.

Urging students to list their sources along the way, perhaps on note cards, may meet with limited success. Here's the good news: electronic means of recording sources have proven to be effective and popular with teens and their teachers.

One such tool is Diigo (*www.diigo.com*), a website which provides a means for bookmarking, highlighting, and sharing the websites discovered and utilized through the use of a downloaded toolbar. Online tools such as Diigo are helping students in their other classes and will, in some form, be available for the long haul throughout our students' academic and professional careers. (Guides to using Diigo can be found online at *www.Teach4SciJourn.org*.)

These e-bookmarking sites are the notecards of the 21st century. When used properly, sites like *diigo.com* prepare students to move easily from the researching to the writing stage because all important material is already highlighted. (This is discussed in more detail in Chapter 10).

Learning to search for information, make judgments about credibility, and keep track of where information is found are the kinds of long-term skills we're advocating; for student journalists, they're also necessary in the short term. Publishable and high quality articles require attributed sources.

From the Editor's Desk: Why Attribute?

Except for accepted facts, ideas, opinions, and theories, science journalists attribute all information included in their articles. For example, "Earth's diameter is just over 12,750 kilometers" does not need to be attributed because it is a

well-accepted number determined by many sources over decades. "There were more than 219,000 new cases of lung cancer diagnosed in the United States in 2009," on the other hand, has to be attributed. In this case, journalists would embed the attribution in the text, writing something such as "according to the National Cancer Institute." A guide on how to incorporate attributions into a news story is provided at *Teach4SciJourn.org*. Here are reasons that attribution is so important for both professional and student writers:

- Attribution provides the reader with important credibility clues to use when making judgments about the trustworthiness of information.

- Attributions give the reader a way to get more information or confirm parts of the article. This is similar to the trail a scientist leaves in a research report about an experiment; the person reading the report should have enough details to repeat the work.

- An article is a historical document, which serves as a record of who said what and when it was said. In fact, it is often said that newspapers are history's first draft.

- Attributions provide protection for journalists by attaching information to individual sources. This is important if information is later proven to be incorrect or deliberately misleading.

- On the internet, attributions often become hyperlinks to other sources. This encourages readers to explore stories in nonlinear ways.

- Both readers and journalists should value multiple sources and perspectives in an article. Attributions allow the reader to easily identify single-source articles.

- Recognizing and writing attributions help introduce students to the wider world. Many students will not have heard of credible sources of information, like the World Health Organization (WHO), Cambridge University, or the American Geophysical Union.

- When students attribute information, teachers can identify plagiarism and respond appropriately. Students plagiarize for various reasons: sometimes to pass information off as their own; other times because they don't understand what they've read well enough to paraphrase. If students attribute information that they've copied word for word, they are probably having a comprehension or a writing problem, not an ethical one.

Lesson Ideas

Credible or Not Credible?

Search-aloud/think-aloud is a good way to demonstrate how you, as an experienced reader, make credibility judgments about websites. You may also choose to do a version of the search-aloud/think-aloud where you deliberately look at sites of varying degrees of credibility on the same topic. Point out the credibility clues you are using as you decide which sites to trust and why.

Here is one story on how it might unfold. At the time we were working on this project, a great topic for unreliable sites was found by searching for "acai berry diets." A Google search on this topic in one class yielded a link to a New York publication—The New York Chronicle. The students examining this website found an impressive array of news logos across the banner—CBS, ABC, CNN, even *Consumer Reports*. The article, *Acai Berry Diet Exposed: Miracle Diet or Scam*, appeared as an unbiased, investigative health report by managing editor Katie Olsen. Most students expressed a degree of confidence in the Chronicle, some even claiming they were readers of the publication.

However, based upon earlier discussions about credibility, the students began asking some tough questions. They noticed the lack of an "About Us" and the tabs across the top of the page—news, weather, and traffic—were all dead, leading nowhere. The health report read more like an advertisement, and upon attempting to exit the site, users were stopped and asked if they were sure they wanted to leave without trying a free sample of the product. A search for managing editor Katie Olsen yielded no one associated with the paper. In fact, the New York Chronicle was purely fiction.

When students later asked about acai berries and what might be the truth about their health effects, the teacher went to WebMD, finding an article written by Kathleen M. Zelman, MPH, RD, LD and reviewed by Louise Chang, MD. A Google search for Kathleen M. Zelman indicated her to be the director of nutrition for WebMD.

Several months later, the teacher showed the class a news article published by the Federal Trade Commission entitled *FTC Seeks to Halt 10 Operators of Fake News Sites from Making Deceptive Claims About Acai Berry Weight Loss Products*. The students wondered what took the FTC so long.

You can find your own "unreliable sites" by, for example, checking health blogs. Interestingly, 16-year-olds seem more ready to believe another 16-year-old who posts online than a medical professional working for a well-regarded hospital. This is another topic to discuss with your class.

For help discussing credibility clues, we have provided annotated web pages in the online supplemental materials on *Teach4SciJourn.org*.

Website Wall

In addition to keeping a record of useful websites in Diigo, a "website wall" can be used, on which students post websites to share with the class (see Photograph 7.1). Strips of oak tag are available for students to use with markers. The website is written in easy-to-see letters along with a brief description (just a few words) and the name of the student posting the site. Not only are student-tested websites available for others to use, discussions arise about the value of the posted sites. As students argue the merits of one website over another, they consider the characteristics of just what it is that makes a website credible and reliable. Before questionable websites are removed, a class discussion can ensue during which websites can be visited and examined together.

Photograph 7.1

Students maintain a running list of credible websites posted in a prominent place in the classroom.

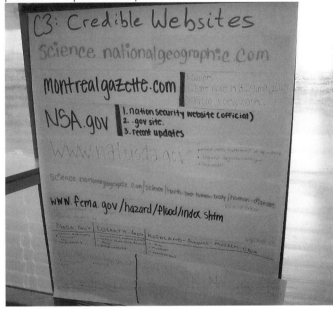

To assess students' understanding of the usefulness and credibility of sites, we often ask them to add a few words that describe the site and why it is credible. For example, in an article about autism the writer added, "…reports Healing Thresholds, a free research-based website directed at families impacted by autism." The readers are then left to judge the credibility of the information attributed to the site and can also trace back to the site to check the credibility using their own credibility clues.

CHAPTER 8

ORIGINAL REPORTING:
INTERVIEWS AND SURVEYS

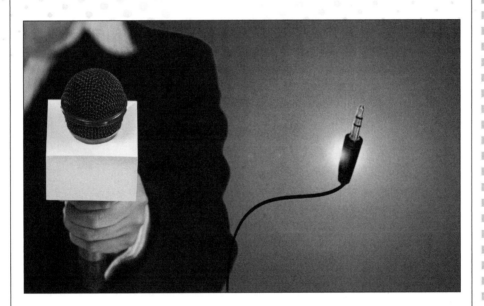

A funny thing happened on the way to helping students write science news articles. When we began this project, we assumed that most of the student work would derive from secondhand accounts, such as easily accessible information on sites like WebMD or in news magazines that describe scientific advances. This chapter, however, is about students coming closer to the sources of information by behaving more like professional journalists. The funny thing is this: The work on interviewing and surveys has opened an unexpected door to science literacy and demonstrated yet another aspect of how science journalism helps students to become scientifically literate.

Thinking of our goal of 15 years from now, we realized that today's students will need to be able to talk with and interview their physician. What are the prescribed drug's side effects? What happens if the surgery fails? How long will I be bedridden? Can you explain that in layman's terms? Where can I get a second opinion? Less scary, but clearly important, the same skill set of asking critical questions is needed to query sales staff about which cell phone or car to purchase. Fifteen years from now, we hope that your students will be prepared to question a member of Congress or read a newspaper and judge the credibility of the experts being quoted. In short, there is a lot of value in learning to talk to experts and asking good questions.

Interviews to Support the News Story

There are three basic sources of interviews that form the foundation of a news story; the reporter might talk to (1) the expert or experts whose research is being discussed; (2) other experts who are likely to have insight into that research but are not personally involved in the work; and (3) stakeholders who have an interest in the research.

In the case of a story on clean water, for instance, likely stakeholders might include the leader of an environmental group supporting new pollution guidelines, politicians, a spokesperson for a corporation charged with polluting the water, and those directly affected, such as anglers or residents along the shoreline. Similarly, a journalist covering the latest research findings on a new medical treatment would query the researchers, but is also likely to call another expert in the field and ask what he or she thinks about this new work. And if the research involved patients, then the journalist may also talk to someone affected by the disease or who participated in the medical trial.

The teachers with whom we work are especially proud that approximately half of our teen journalists now include original interviews in their stories. Admittedly, many of the interviews are with people whom the teens know well—the aunt with diabetes, the mom who is a nurse, the grandfather who quit smoking.

But, as a teacher, you can usefully think of these interviewees as sitting on a continuum. On one side of the line are those people the students know well, for instance, a parent or grandparent; some writers even use their own stories as a springboard for their article. On the other side of the line sits the PhD or MD researcher who has published papers in scientifically well-regarded journals.

Pushing young journalists along that continuum is a worthy goal. Actually, helping students realize that there *is* a continuum is helpful as they seek to challenge themselves. *SciJourner.org* articles have included interviews with the local pharmacist, a wildlife biologist, a tattoo artist, a dietician, and even an EPA official. What we can say for sure is that interviews make for livelier stories and those stories are more likely to get published.

The challenge is that many young people get very little practice conducting interviews or even talking with adults. Even if they don't believe that adults are inherently smarter or more powerful creatures, teens are often uncomfortable querying adults. Many are downright afraid to talk to grown-ups. They don't know how to approach them, how to ask difficult questions politely and persistently, or how to follow up once the interviewee has said something of interest. These are skills well worth teaching from a science perspective.

Contacts

How do science journalists locate contacts? The process is often the same way responsible adults find a doctor. It begins with someone the reporter knows and trusts. It could be a scientist, high-ranking government official, media officer for a company or organization, or a name on a press release from a credible group. Journalists will often ask a source, "Who should I talk to?" The point is that the journalist does not begin with a random

Photograph 8.1

SciJourner reporter Damonte Johnson conducts an interview with Saint Louis Science Center employee Jennifer Jovanovic.

Photo: KiOntey Turner

search through the phone book. The reporter targets credible sources, sources most likely to answer questions related to or raised by the subject at hand.

In schools we may have similar access to expertise. If each member of the class knows what their classmates are looking for, they may be able to help locate an expert. The school nurse might talk to a student about athlete's foot. The coach could address questions for a sports science story. A classmate might have a parent who is a wildlife biologist or a civil engineer.

Some students seem unduly impressed by credentials. One young man said he didn't know anyone to talk to about bridges, although his father had been a construction worker for years. "He doesn't even have a college degree," our young journalist told his teacher.

It is important to remind students that most local journalists rely largely on "people in the neighborhood." As an assignment, have students look at the local TV or newspaper press and list where the interviewed experts come from. The owner of a store that was recently flooded, the pipe fitter, and the county water system engineer are all expert sources.

What about working scientists? Not surprisingly, a science researcher busy in the lab responds differently to a call from the New York Times or *Science* than she would to a student writing from Fairville High. Local colleges are more likely to respond to student queries than the college three states away, and the local professor may well have a friend at another institution who could offer advice.

Help students begin to think aggressively about expertise and how to access it.

- Counties and states also have agencies staffed by folks with expertise, such as farm extension offices, departments of natural resources, and local waste treatment facilities.

- Professional societies, such as the local chapter of the Academy of Sciences, American Chemical Society, American Geophysical Union, or National Society of Black Engineers, are often willing to help out an interested student. Many medical schools also have young physicians groups that are interested in outreach to the community.

- Large organizations or companies often offer the services of a public information officer (PIO)—what most of us think of as the PR (public relations) people. It is their job to match reporters with the right person in their "shop." In our experience, some PIOs are incredibly helpful and some will not give you the time of day. (Professional journalists have the same experience; it is not limited to high school students.) The best way to get results is to appear professional. And being aggressive is part of the job; keep at it until you get an answer.

First Steps

The word *interview* covers a wide variety of encounters, everything from simple fact-checking to multipart conversations that allow a journalist to think about and do additional research regarding the topic in between interview sessions. As adults we understand that the preparation for the simple fact-checking phone call is significantly less than for the full-blown interview. However, many young people get trapped at the opening gate. "They are all adults and I don't particularly want to talk to them!"

Here are some suggestions designed to help students build their confidence as interviewers.

1. Ask students to read over the interviews on *SciJourner.org* and imagine the context in which those interviews took place. Chat together about who they see as more difficult and easier people to interview. Why?

2. Role playing in class can also be useful. Students can be paired to talk to one another, but if an expert is needed, have the teacher play the interviewee or invite in a colleague; perhaps the principal or science supervisor.

3. SciJourn teacher Déanna Scipio suggested asking your own adult friends and relatives if they might allow students to call them from school in order to practice.

4. Depending on the anxiety level of your students, be an easy or a more challenging interviewee (or ask your colleagues to be more or less difficult). If possible, do these role-plays in a "fish bowl" and analyze the moves the interviewer makes in order to get the expert to talk either during or after the interview. Remember, unpacking the role play is at least as important as the role-play itself.

Although these are all useful suggestions, authentic confidence comes finally from practice in the real world.

Different Kinds of Interviews

It is not so much that one sort of interview is easier than another in a global sense; the difficulty is personal. Calling to check on the spelling of a name seems like "a piece of cake." However, for the student who has already called twice and lost the piece of paper on which they wrote the correct spelling, this is an exercise in humiliation.

Fear of being embarrassed actually ranks high in the interview worry list for many teens, and even some adults. What might embarrass one teen (or adult) seems absolutely comfortable to another. There are families where talk of disease, for instance, is considered something akin to bad luck, while other families are comfortable in passing information on to a younger generation. From a science literacy perspective, this becomes an important issue. Medical histories allow us to make well-informed decisions about our own health and that of our relatives. And yet pushing students to violate family values is surely problematic. There is, as yet, no best practice.

A group interview is one exceptionally useful way to get students started. Local colleges are gold mines for this activity; many professors are eager to encourage science interest among high school students. In one SciJourn classroom, a professor of physics conducted a "press conference" on new car technologies. After a 20-minute presentation, the class, prepped to act as journalists, bombarded him with questions. Press conferences are also available online, although they rarely allow for real-time questions and answers.

Another easy start is to interview someone close to home. An aunt, a family friend, or a neighbor might be the start of a story. Many of the *SciJourner* health and medicine stories begin with a family member or neighbor dealing with a disease. They can also be the source of "man-on-the-street responses" that work in some ways like a kind "people's" peer review. For example, a student writing a story on healthy meals might question family members and friends, "When you buy fast food, do you consider the calorie and fat content?"

With added confidence, students can "graduate" to interviews that help them learn about authentic scientific practices. This works especially well if students visit a lab. Note, however, that it is exceptionally difficult to develop questions about scientific practices when the instrumentation or mathematics is far beyond the understanding of the interviewer. Journalists understand this and have developed a long list of general questions to guide them through nearly all interview situations. Questions about stumbling blocks or surprises the scientists encountered in their research, potential applications and their probable cost, challenges to commercialization or trials with humans, or next steps in the research are pretty standard. From the "15 years later" perspective, it anticipates what we hope to be the automatic response of your students years from now to something like a drug prescription—"What are the side effects?"

Group interviews in the classroom are another way to query scientists. It has the benefit of spreading the burden of asking questions around the room. We already mentioned the class "press conference" as one approach, although it can be dominated by a few voices and it relies on spontaneous questions. Moreover, a speaker might run over their allotted press conference time, leaving you less time for questions.

An alternative approach is to again bring a scientist to the classroom visit, but research his or her work ahead of time. Nearly all researchers post web pages highlighting their work and interests, which is a good jumping off point for some online searching. With this background information, divide the class into teams of students who work together to formulate and write down a set number of questions, along the lines described in the next section of this chapter. The questions are then compiled, ranked, and used to query the visiting scientist. Instead of a press conference, the entire session is taken up with a question-and-answer period.

We have used this latter approach during our SciJourn PD sessions, with great success. Although everyone participates in drafting the questions, only one person asks them. Interestingly, this is the same approach followed by the National Press Club in Washington, D.C. It maximizes the time because questions are not repeated. Annotated videos of this type of group interview can be found on *Teach4SciJourn.org*.

Preparing for an Individual Interview

A teacher at a recent professional development (PD) session suggested an article focused on whether rooms filled with salt, known as halotherapy, really help individuals who suffer from asthma, allergies, and sinus problems. The website for the salt room promises relief from symptoms for 3 to 12 months.

The topic meets our standard I—the salt room is local and the topic is narrow and presents an interesting angle. But who might be good to talk to about this topic? For this example, it would be a smart exercise to generate and prioritize a list of potential interviewees with students. Remember, all student journalists, like all professionals, need not agree. A more structured approach would be something like Figure 8.1.

Teach4scijourn.org also has a lesson designed to help students think about who might be appropriate experts to call. If you, the teacher, are working on an article in advance of your students, consider your own topic as a place to begin this sort of brainstorming.

Professional science writers prepare for an interview by looking for background information on the web, reading the appropriate article or press release, and talking to background sources (which could be their science teacher). They want to be well-informed so that they can intelligently engage in discussions with the scientists who conducted the study or those they are asking to confirm or critique the research. It takes the journalist a step beyond those general questions. For example, how much a new drug will cost will get a number, but the prepared reporter might ask about whether it is better than a cheaper drug that is already on the market. While students worry about being personally embarrassed, journalists worry that they may not know enough to ask good questions. This is a good worry to share with students!

What journalists do best is ask questions that help explain an issue, whether talking face-to-face, over the phone, or via e-mail. Asking the "right questions of the right people" is a skill that reporters covet. Good journalists exercise great care in who they interview and the kinds of questions they ask. Interestingly, the end result is that two or three journalists covering the same story can end up with different articles, depending on whom they interview. However, each article is probably closer to the truth than if the story had been written as a press release, only citing one source.

It is important that students have a clear purpose in mind when they call a potential interviewee. We encourage students to act professionally and identify themselves as journalists. Not all queries are going to be answered, but *Sci-Journer* writers have had successes. In some cases, the source responded because there was a connection (the family doctor or a local college) or they were easily available at an event, but in other cases it has been a student reporter not willing to give up. And sometimes no answer is an answer in itself. Journalists will say

---

Figure 8.1

A Lesson Plan on Who to Interview

From the list below, select two to three experts to interview for this story. Circle the experts you would choose. If they are not available to answer questions, who would you talk to next?

- Asthma sufferer who has used the salt room and been pleased
- Asthma sufferer who has used the salt room and not been pleased
- Asthma sufferer who has never heard of the salt room
- Allergist
- Financial backer of salt rooms
- Gerontologist
- Official spokesperson for salt room firm
- My personal physician
- Someone in charge of the salt room firm
- Secretary who makes appointments for salt rooms
- Worker in salt room building

Pick two of your circled experts. Beneath each, write five questions you would ask that person about the topic of salt rooms.

that Mr. X or company Y did not respond to our inquiries, leaving it up to the reader to decide if they are avoiding public comment for a reason.

Developing and arranging questions to pose is somewhat of an art. Here is advice to get you started.

- *Do your research!* Be prepared. Refer to your research. It will give you credibility.

- *Write down and rank your questions.* Place the most important questions at the top of the list in case time is limited. Avoid "yes or no" questions, if possible. (One exception: It may help to confirm something with the expert—"Is it true that the batteries on electric cars need to be replaced every two years and are expensive?" The reporter can always follow that with a more open-ended question— "What new technologies might avoid this problem?") The more the reporter gets the interviewee to talk, the better!

- *Avoid simple questions.* Does it make sense to ask a medical professional: "What is diabetes?" when the answer is easy to find on credible sites? A better question might be: "Are there any new treatments for type 2 diabetes?" or "Why are there so few new treatments for type 1 diabetes?" That shows that the reporter has done their homework and is looking for something new. Experts appreciate reporters who ask smart, tough questions and, as a result, tend to talk more freely.

- *Think carefully about the phrasing of a question.* "Why do people buy hybrid cars?" is not as good as "Why would someone buy a hybrid car when they can buy a regular car with really good gas mileage for less?" Once again, good research ahead of time makes a difference.

An enlightening exercise involves analyzing a prerecorded interview—we have used an NPR interview on bees for this purpose (a link is provided at *Teach4Scijourn.org*). We begin by telling the students: who was interviewed, the topic, and when the interview took place. We then give them a bit of time to research the topic. Next they create their own questions and, as a class, compile and rank their top five or so. We then play the interview and compare our list to the reporter's queries, stopping occasionally to discuss and critique the professional's questions.

Conducting the Interview

The interview itself will vary depending on setting and time allowed. The best interviews are face-to-face. Body language and voice tone really help in understanding an answer. Phone interviews are next best; voice intonation provides clues. E-mail interviews are OK, but they tend to lack the spontaneity of the immediate response and nuance of voice.

- *No matter how the interview is conducted, identify yourself as a journalist.* We can't reiterate this one too many times. By labeling himself or herself as a journalist, the student reporter becomes a credible professional, and whatever is said is "on the record."

- *Establish a meeting time for the face-to-face or phone interview.* Ensure that there will be enough time for the conversation; often 30 minutes is a good length. Be on time for the interview and dress appropriately. Don't be surprised if the interview runs longer than the allotted time. Experts love talking about their work, especially to someone who is interested and prepared with good questions. If the source cancels, ask for another time and don't be discouraged.

- *Either audio record the interview or take notes.* As a novice, it can be challenging to listen to the interviewee's answers, take notes, and remember your next question. Audio recording reduces the pressure. Practice with the equipment ahead of time and make sure to install a fresh battery—nothing worse than returning home only to find no recording. Ask permission to record. Be sure to get the spelling of the interviewee's name; no matter how simple it may sound. A John Smith might spell his name as Jon Smyth. Also ask their official job title.

- *Get a direct quote.* Besides the information provided, it is important to get one or two direct quotes from your source for the story. Journalists develop an "ear" for good quotes, sometimes asking questions designed to elicit them—"What would you compare this to?" "How might you describe this to a non-scientist?" "Were you surprised by these results?"

- *If you e-mail the questions, keep the list short.* Typically, five or six questions are all anyone wants to answer. Again, choose your best open-ended questions. You might also include the "anything else" query described below. And check your spelling and grammar. Finally, use an e-mail address that says "I'm professional."

Photograph 8.2

The *SciJourner* press pass, such as the one shown here, has been used by student reporters. Student journalists borrow it when they go out on assignment to conduct interviews or engage in additional research. Asked if other students were impressed by the press pass, one author said, "I don't care if they are. I am!" When student journalists become published authors, they receive their own press pass to keep.

Photo: Shannon Briner

Interview planning tools are available at *Teach4SciJourn.org*. Here are a few basic techniques, journalists (you and your students) may find helpful:

- *If the questions are personal rather than professional, let the source tell their story.* Interviewing someone who has a serious disease or experienced a terrible accident can be intimidating, but it is frankly therapeutic for the person. It may be easier to just let them talk, then ask questions. Be sure to get those personal details, such as how it felt, how they struggled through it and how they are coping with it now.

- *Don't be afraid to challenge the interviewee.* If the answer raises alarm bells or seems to make no sense, then don't let it go by. "Do you really mean to say X?" "If there is no difference between bottled and tap water, then why do so many teens buy bottled water?" "Do you really expect teens to sleep 10 hours a night, even if it is good for them?" "If bacteria are all over the kitchen counter, then why isn't everyone sick all the time?" Your allegiance is to the reader, not the source. You need to answer the questions that they will ask when they read your story.

- *Listen to the interviewee's answers.* The student reporter might want to ask a question *not* on their list if something surprising or unexpected is said, or, as stated previously, challenge some answer. Again, audio recording makes it easier to concentrate on the answers.

- *Do not worry about quiet, sometimes awkward moments between questions.* Actually, by waiting a moment or two to follow up on an answer, the interviewee may feel the need to keep talking—and especially interesting comments might emerge.

- *At the end of your interview, always ask, "Is there anything else you would like to say that I did not ask about?"* This is the journalist's secret weapon; sometimes the answer to that last question is quite useful in your story. And always thank your interviewee for their time!

To learn more about interviewing, consider inviting a working journalist into your class. Science and health reporters are, sadly, getting harder to find, but newspapers and local TV stations that still have them on staff will usually jump at the chance to talk to your class.

Protect Your Sources

We hope that no one interviewed by a student asks for anonymity (to not be named). We typically discourage getting into that position; it compromises the credibility of the story by making the sources "invisible." The one time that we did encounter a source asking not to be named in the SciJourn project, we advised the teen not to use that source.

However, comments from students under the age of 18 fall into a special category. Often, they are cited as Rob, age 15, or Latisha, a high school sophomore. The absence of a last name is a form of protection. We assume that after 18, people are adults and responsible for their statements.

After the Interview: Deciding What to Use

Ideally, there is at least one good direct quote from the interview for the story. It shows that the author did indeed talk to the source. Moreover, a good quote should reveal something about the source. Typically, it is something lively, insightful, clear, and perhaps surprising. Have students collect intriguing quotes and discuss with them why these quotes were useful. What insight did they offer the reader?

Here are some *SciJourner.org* favorite quotes:

- "Not washing your jeans? Ewww," exclaimed one freshman. "They'd be so rank" (Honiker 2011). The story centers on the environmental advantages of not washing blue jeans.

- "Hunting is necessary to keep the deer herd at a healthy level," Missouri Department Conservation wildlife biologist John Vogel tells *SciJourner.* "Without hunting, you could see a variety of effects ranging from a largely diseased deer herd to deer running all over the road and getting hit by vehicles" (Tappmeyer 2010).

- "My personal opinion is that we will be able to meet the challenge of feeding the growing population. It's always easier to solve a problem you've identified early," writes Weiss (Guthrie 2011). The story on disappearing farmland quotes Jacqueline Weiss, a Scientific Applications Specialist with the biotechnology firm Mosanto.

Rewriting a direct quote is discouraged. If there are changes to the direct quote, it is marked with square brackets. For example, the expert said during an interview: "Researchers believe that the calcium channel receptors have something to do with it and that some blood vessels are more susceptible to these spasms, which would explain why migraines can run in the family." However, the quote is too much jargon, and what is "it"? The revised quote read: "Researchers believe that the calcium channel receptors [which regulate the flow of calcium ions across the membrane in all cells] have something to do with [migraines] and that some blood vessels are more susceptible to these spasms, which would explain why migraines can run in the family." (Ragland 2010)

If the comments are too technical, awkward, or long-winded to "fix," they get translated and presented as indirect quotes. However, quotes, as well as descriptions born from interviews, must stay true to the source's thoughts.

Surveys

While an interview elicits individual responses to a question and enlivens an article, a survey seeks to gather information from a wide swath of individuals. It was a SciJourn student that first clued us in to the power of social media and surveys. For a story, she simply queried all of her Facebook friends on an issue and, in no time, had over 50 responses. Other students have used *ballot-box.net*, Google docs, and paper forms. All have quickly gathered large numbers of responses.

In our view, gathering information from original surveys is the same as original reporting. A story on green schools gets a fresh angle by surveying students in the school on whether they think their school is doing enough to be eco-friendly. Or a story on sleep gets a fresh look by including a survey of students' sleep patterns.

We have found that many students really enjoy conducting a survey. What they do with the data is another issue. Over and over again, we find that many young people are stumped with what to do with original data, and many miscalculate percentages or try to work with statistically insignificant subsets (e.g., in the 50 respondents, only 2 are junior-year females and 1 says yes, which is reported as a 50% response rate). We strongly urge you to work closely with the survey takers.

Some Survey Guidelines

1. Keep the number of questions small, typically five or six is the maximum. No one likes filling out a 10-page questionnaire, and many won't.

2. Think carefully about how you are going to sort the data. Age and gender are common questions, but for a survey of football injuries or computers in the home, those categories may make no sense.

3. Be prepared to "collapse" some categories. The sleeping habits survey had asked for age, but the age range was too small to be meaningful. The author just said that the responses were students between 15 and 17.

4. When writing up the survey for the story, it is important to explain how the survey was conducted, exactly how many people were polled, and when the poll was taken. Yes, this is not a true scientific survey, but it is data and should be reported as data.

5. Survey results are typically anonymous; however, it is always good to interview one or two of the respondents for direct quotes.

6. Make sure to be aware of school policies. *SciJourner.org* reporters have surveyed teens on drug use, but through a Science Center. Many schools have policies that forbid asking these types of questions; the

concern is that names may accidently be released, setting the school up for a lawsuit.

Students tend to be excited about conducting surveys and interviews, a fact that surprised us intellectually. This kind of original reporting is also a good approach for reluctant writers. One *SciJourner* teen, whose writing skills were weak, put all of her energy into just collecting original data on unwed mothers. It turns out that she was also interested in graphic arts, so she spent her time working on ways to present the information visually. In the end, she created a chart with graphical information and citations to credible sources and followed it up with a movie that included interviews with three teen moms. Chapter 12: Beyond Words discusses other formats for presenting information from surveys and interviews.

References

Guthrie, K. March 25, 2011. Expanding population eating away at farmland. *www.SciJourner.org*

Honiker, K. July 27, 2011. Wanna save the world? Don't wash your jeans. *www.SciJourner.org*

Ragland, N. July 27, 2010. What's wrong with my head? *www.SciJourner.org*

Tappmeyer, B. February 26, 2010. Hunting proves both profitable and crucial. *www.SciJourner.org*

CHAPTER 9

CHANNELING YOUR INNER SCIENCE TEACHER:
CONSIDERING CONTEXT AND ACCURACY

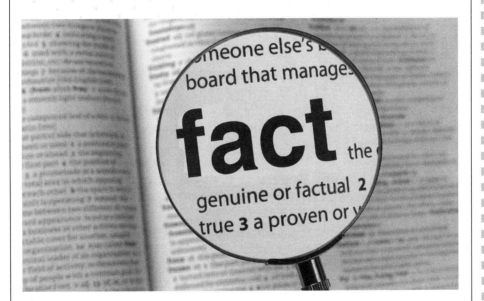

This chapter includes ideas for helping students put information into context (Standard IV) and fact-checking their own and classmates' articles (Standard V). Up to this point, as we discussed the creation of a news story, you as a science teacher may have felt you were on mushy ground. In considering science context and accuracy, however, you may well feel that you have returned to terra firma. Once students have their idea for a story in place and have done the basic research, it is time to put the targeted and specific ideas into context and to check to make sure that the details are up-to-date and as correct as they can be. The act of contextualizing and commenting

on the accuracy of information about technologies and discoveries; differentiating between the widely accepted and the emergent; attending to the nature, limits, and risks of a discovery; and integrating information into broader policy and lifestyle choices is, in many ways, familiar to teachers. It is what has made science not only important, but interesting.

To ensure accuracy and better understand context, it is necessary to do additional searching and reading, some of which is likely not to become part of the article. Teachers should not expect to be entirely responsible for knowing enough to "correct" and provide guidance about each and every topic chosen by students; that would an overwhelming—actually impossible—task. What we can expect, however, is that teachers can consciously name and appreciate the act of trying to get it right.

Our editor reminds us that we actually *can't* catch every mistake or "get it completely right." Papers print retractions and corrections every day. Perfection isn't possible, but instilling certain habits of mind in students is. Remember, that's our goal. Those habits of mind are the keystone of science literacy. The difficult task from a teacher's point of view may be holding back, sorting through what details matter, and getting the students themselves (not you) to do the work.

Why Should I Care?

Context is about fitting one little story—the kind that can be covered in 500 words—into the world of science-at-large. Why does the topic chosen for this story, a topic which has been condensed in 500 words, really matter? Is it a big deal? Is it an interesting example? Is it an embryonic idea that if true shakes up older ideas? Does this story tell us something about how science as an enterprise moves forward? The trick for the author of a news story is to be able to answer how the little idea fits into the big idea in just a few words. It's sort of like the "elevator speech" about why this story is worth reading and why the reader should care.

Students who are trained in science fair report writing sometimes believe that context is equivalent to historical background. Consequently, they may produce an article about landfill gas being used as an energy source—a very cool idea—that includes a paragraph about how the ancient Persians pioneered the technology. It is an interesting point, but says little about why engineers are so keen on the concept at this point in history. The effort to put a story into context will feel especially familiar to those who have attended to efforts to build science, technology, and society (STS) issues into the curriculum.

In our experience, students don't include sufficient context and don't have a good feel for why context matters. A young author needs to see that the article she is writing about her brother's autism—and her interest in her brother's autism—will benefit from knowing how many Americans are estimated to be autistic. Context is about what's there, knowing what numbers matter and whose work represents the current standard, but it is also about what's missing.

CHANNELING YOUR INNER
SCIENCE TEACHER: CONSIDERING
CONTEXT AND ACCURACY

Front-Page
Science

CH. 9

When we think about these students years after graduation, their difficulty with context is concerning. If my child is diagnosed with leukemia, I will need to navigate a maze of medical issues: which treatments show promise and which are experimental, which hospitals and doctors are best, and how to comfort and care for the patient. That means that I actually should care if my child's disease affects many people because common maladies are typically better studied and often produce more active patient support groups to help families cope and lobby Congress for money.

Context has to do with how the topic under consideration might impact our societal or ecological system and how those systems potentially impact the science being discussed. Beyond that, students must weigh the credibility of the evidence, not an easy task for even for the most scientifically literate among us. Indeed, what is believed today (a new discovery or treatment) may prove to be not true in the long run, which is part of the reason why all science needs to be viewed in context.

For journalists, it is almost formulaic: A good article needs a paragraph or two that answers questions like "Why should I care? Why is anyone doing this? Is it better, cheaper, faster, or smarter than what came before?" The goal of the context paragraph is to show that science has broad consequences, is an ongoing enterprise, and is part of a continuum of investigation and learning. Scientifically literate adults demand answers to context questions about public policy issues. When a politician proposes a new bridge, for example, citizens would do well to know the environmental impact of the structure and the cost, and not just that it will help with traffic problems.

Technology stories typically call for explanations of what came earlier, how much the new technology will cost to build and buy, why it is better, and when it will be available. Environmental stories might address how the findings potentially affect legislation, industries, or practices; who and how many people, species, or ecosystems are at risk; and how this fits into other observations or data.

In medical stories, context is pretty straightforward. How many people are diagnosed with this disease each year? If it is potentially serious, how many die or are seriously incapacitated? When reporting on a new finding, the journalist needs to describe how scientists explained or diagnosed this disease previously. Is this an accepted finding, preliminary, or a view held by just a few?

Look at the context paragraphs in stories. Here is a student example of a job well done: "The March of Dimes states that Down syndrome affects about 1 in 800—about 3,400 babies a year in the United States. According to the National Down Syndrome Society, there are more than 400,000 people with the disability in the United States" (Boyd 2011).

As usual, we suggest beginning by having the teacher model her processes. Think aloud about an article, stopping before you get to the context paragraph(s). Work with students to brainstorm what they think should be included in the context paragraph. Or you might introduce context by simply

removing the context paragraphs and asking them either individually or in small groups to create the context. They would first need to identify what kinds of information they want to include and then find credible sources that provide that information. If several groups are working on the same article, you can see what contextual information most students agree is essential and what information they view as optional. Remind students that within a short science news article they can't possibly include all the contextual information that is available and should provide just enough so that the new information makes sense.

During any read-aloud, students can be encouraged to identify the context paragraphs. Do students recognize when it's missing? Better yet, see if they can guess what questions will be answered in the context section.

Finding contextual information calls for some extra research; look specifically at the search star (Figure 7.3, p. 97) if students need ideas. It also requires that students decide what is most important—a critical-thinking skill, to be sure. In most stories, the essential contextual information is the *number*—a percentage, cost, estimate, distance. It is hard to write a good story without a number.

Sometimes, it is the number that needs context, especially when that number is really big or small. In a story looking at the loss of farms in the United States, the author found a number for the area of the country and a second value for total U.S. farmland. However, these numbers only made sense to readers when she turned her information about the amount of farmland into a percentage of the entire U.S. landmass. Oddly, the author was uncertain as to whether she could include the percentage in her story because the percentage itself was not found in an online source or textbook!

If your students are not used to thinking about context let them practice. One teacher suggests beginning SciJourn work with ferreting out context information rather than finding a topic to write about. In this case the teacher names the topic and says to students "Tell me why it matters." This was a good way to teach both context-related searching and why certain issues are worth studying.

If curriculum is not your main focus, try beginning with health issues. Students can certainly find context information about a health condition affecting someone in their family. Sometimes context has to do with political or economic ramifications. Finding those contextual details often involves understanding credible sources on a different level. Does a political party or partisan website have more credibility than CNN or NBC?

Accuracy: Just the Facts, Ma'am

From a scientific perspective, the word *fact* is problematic. Scientists value observations, theories, and hypotheses, but for them a fact implies something like absolute truth and that, we know, is not to be had. In ordinary speech, however, facts are what the old *Dragnet* cops looked for—bits of information that can be confirmed. Generally, we think of a fact as something that is believed to be true. However, among scientists, the term *belief* is also problematic. Does it matter

CHANNELING YOUR INNER
SCIENCE TEACHER: CONSIDERING
CONTEXT AND ACCURACY

Front-Page
Science

CH. 9

what I believe? Isn't what I believe actually my opinion? Once astronomers believed that the Milky Way was the only galaxy, then Edwin Hubble peered deep into universe and showed us that we lived in one of countless galaxies.

What scientists do is make *assertions* based upon the *evidence*. Assertions and evidence are on shifting sands—subject to change and revision when new data are uncovered. It is in this arena that student journalists must determine the accuracy of facts and assertions based upon the available evidence. The linking of science with journalism can make the use of fact and assertion interchangeable. It's also important to realize that some theories and assertions are foundational and agreed upon, while others are new and still very much under review.

Newton's Laws, for instance, are well-established, what we could call axiomatic. Research on climate change has, in our lifetimes, gone from questionable to well accepted, and the doubters' arguments have moved as well. Other scientific theories have lost credibility, such as "cold fusion in a jar" and Hoyle's steady state expansion of the universe as an alternative to the "big bang." Meanwhile, other theories get new life—epigenetics is an emerging field of study that finds that environment does influence some aspects of inheritance, echoing Lamarck's early 19th-century hypothesis that was discarded in favor of Darwinism.

"I found the writing not to be hard, but I wanted to know everything before I wrote, so I just kept searching for answers."
—Colleen Abbott

Knowing the status of theory-generation or research is an important part of fact-checking. It also may be the most difficult part; understanding whether research is well established or embryonic is absolutely key to "getting the facts right." Most teachers think of climate change as the obvious example (and it is a good example), but also consider health-related issues. What do we know, for instance, about taking ginkgo biloba to stem the tide of memory loss?

When we have asked teachers to respond to sample student articles, practitioners have jumped all over inaccuracies—some tiny like the capitalization of a chemical symbol and some huge like a confused understanding of a concept or process. Given our idea that science literacy is about the skills needed to navigate science material 15 years after graduation, how important is an inaccuracy here and there in a science article submitted in a high school class?

Let's think for a moment about those students, now all grown up. Can we help develop their ability to watch or read a report about a new technology and recognize when something just doesn't "sound right"? Will they be able to listen to their elected officials discuss energy policy and recognize information that needs to be double-checked? When the next health scare affecting their children arises, will they have the resources to determine if it's a big deal or something they need not worry about?

Learning to Be Accurate

Many literacy teachers have encouraged their students to interpret and assess text. *Does this text make sense, given what I know? How do I know if this is or is not right?* Teachers ask themselves, "If this were the only text students read, what would their perception of the problem be?"

We can all remember reading something that raised red flags in our minds, but what raises flags for one person might not for another. Our professional science editor, with years of experience reading science news and a PhD in chemistry, surely has a more finely honed sense of when something is inaccurate in a science news article than our high school English teachers. Our students may be better at noticing illogical or inaccurate statements in certain articles—those about video games or other new technologies, perhaps—than some of their teachers.

We return to the idea of science literacy as distributed, the property of a community not an individual. In a classroom, this distributed community can come to life during fact-checking exercises. Conversations about what doesn't sound right—and why—in different articles can help the entire classroom improve their ability to ask questions and check information.

Think of fact-checking in this sense as a form of estimation. We get better at estimating distance or time by doing it. The same thing will happen with fact-checking—ask students "Is there anything that doesn't look right in this article? Anything you don't understand as a reader that you need to?" These are places to check on the information provided.

Through read-alouds we can practice fact-checking. Take an article and omit or change information to see if anyone in your class catches what's wrong. Help students develop that "critical eye."

As students come to assess assertions and attend to how reasonable those assertions may be, they become more critical readers—they gain what is sometimes described as the *gut feeling* scientifically literate people acquire about the things they read and hear. And they learn that this gut feeling needs to be checked. Assertions in science journalism must be supported by evidence, ideally through the attributions that show the trail back to the author's sources. We have seen how important attributions are to credibility (see Chapter 7). When facts and assertions are questionable, it is time to evaluate the source of the information and its accuracy. We ask our students to write detailed lab reports so the experiment's results can be *replicated*; likewise, we ask them to write news stories with enough facts and attributions that another person can follow the trail back to its source and confirm the story.

Even when appropriate attributions are present, fact-checking may still require a degree of prior knowledge on any given topic. An article that identifies Hannibal, Missouri, as being located on the banks of the Amazon River would obviously be inaccurate regardless of the attribution. But if the article identified Hannibal, Missouri, as being located along the Missouri River, readers might

not question or check its accuracy. (Hannibal is actually located along the Mississippi River.)

Part of what we do as teachers and adults is share background information. An adolescent who grows up eating at the dinner table with two PhD scientists comes to school with a different level of background knowledge than a student whose efforts have been spent helping his parents sell groceries in the family store. It doesn't mean that the merchant's son doesn't know much science. What it does mean is that this young man needs to gather information differently than the kid who lives the science talk each and every day. One function of schools is to help build such general knowledge and level the playing field. But remember, when it comes to an article in which teen knowledge is better than our own, help students appreciate their own gut feelings. For example, does it make sense that only 20% of students had sex edcation in high school? In this case, they will have a better idea of what makes sense than an individual deemed scientifically literate who spends 12 hours a day in the lab!

Fact-Checking

The term *fact-checking* comes from the journalism community and is fundamentally about accuracy. One or two inaccurate statements in an article can undermine the credibility of the entire story. Making sure that the information is accurate and contextualized means getting the science "right" as a means of maintaining credibility.

The process of fact-checking is also a call for "attention to detail." Is it the National Institute of Health or the National Institutes of Health? (It is the latter.) Is it Jim Wazkowski or Jim Wazkowsky? All details go under the microscope in journalism. In fact, in journalism classes, students fail the assignment if they misspell any names!

This level of detail mirrors what is expected of scientists. The compound is not red, but rather reflects light at 710 nanometers. It isn't skin cancer, but rather basal cell carcinoma. The terms and numbers matter. The kind of accuracy called for is important for several reasons. First, it goes to the issue of getting it right, showing the author did her homework and understands what the community she is writing about values. Second, accuracy goes to the issue of *replicability*—the ability to follow the processes that the researcher, or in this case the author, engaged in. Third, because the article becomes part of the historical record and we want people in the future to be able to count on what has been said, accuracy matters. Scientifically accurate language also enables the writer and reader to search for additional information and find credible sources.

Although it is our experience that teachers care about accuracy, journalists take the business of accuracy and fact-checking a few steps further. Use the following questions as a guide as you and your students fact-check articles.

Is the Information Right?

"It is interesting how we can often see the issues in others' work, but not as much in our own work, even when it is the same problem."
—Colleen Abbott

Use your instincts. Sense problems. If it sounds wrong or strange, it may very well be wrong. You then need to check your instincts. One of the biggest challenges of the SciJourn project is that there is no answer book. You and your students will often be co-learners. Credible information sites, such as Wikipedia or WebMD, are great places to quickly check science facts. Asking the author to explain the science offers another clue. If he can't articulate it, then red flags should go up. (This also becomes a check for plagiarism.)

What about the English or Journalism teacher? If the teacher can't understand the explanation, then call it into question. The author may be equally confused. It also helps to build alliances; can you run the explanation past the chemistry or physics teacher? They should be able to help you check.

When errors are encountered, put the job of correcting the problem back onto the student. "Really?" we might ask, "Please recheck those facts." "This doesn't read right, please recheck your source."

Numbers and statistics are places where teens regularly trip up. A story on the incidence of diabetes among immigrant populations originally said that "1.1 million immigrants developed type 2 diabetes." However, when we fact-checked the source, the source actually said that out of 1.1 million immigrants, some populations showed a two- or threefold higher risk of developing type 2 diabetes than the native population. The percentage risk had been transformed and huge numbers of immigrants were incorrectly described as diabetic.

Look also for contradictions in text. Finding such problems, especially when they are paragraphs apart, is a real content-reading skill. For example, a story about the serious pollution problem in the Mississippi River included this interesting statement: "In 2004, a study carried out by the Louisiana Department of Environmental Quality tested fish tissues for more than 100 toxic chemicals. The results showed that 95% of the toxic chemicals did not show up in the data, and any toxins detected fell below the FDA standards for an edible fish." This information effectively undermined the author's story. Students avoid this problem when they learn to be critical content readers of their own work and not just focus on spelling or grammar mistakes.

Is the Language Right?

Journalism and English teachers get this one. We often see a statement such as "Researchers *proved* ..." or "Scientists *know*...." In actuality, researchers *show*

CHANNELING YOUR INNER
SCIENCE TEACHER: CONSIDERING
CONTEXT AND ACCURACY

𝔉ront-𝔓age
Science

CH. 9

and scientists *suspect*. Fixing these mistakes is more than just editing; these terms have important meaning in science. We realize that plants do not *eat* and astronauts do not *fly* through the space station, but such terms often go unnoticed in the popular press. Attending to language alerts students to ideas about the nature of science. In addition, it gives them important clues about how carefully other reporters have attended to science concerns.

Sometimes, careless language reflects a lack of knowledge about how information flows. One student wrote: "United Healthcare *found* that salt disrupts the organisms' ability to pass fluids in and out of their body (osmoregulation)." Because United Healthcare is an insurance company, it is highly unlikely they were the group that "found this out."

In an article that was eventually published in our magazine, a student initially wrote that drunk driving was "the leading cause of death" for teenagers. This raised a red flag for us; when we checked the reference it turned out that drunk driving was described as "a leading cause of death." The difference between *a* and *the* might seem minor to a student, but understanding this kind of contextual detail is an important science literacy skill.

In that same article, the student also claimed that the "majority" of teenagers who get into vehicle crashes had elevated blood alcohol content levels. Once again, we checked the reference. Speed was listed as the number one cause of vehicle crashes involving teenage drivers, not alcohol. We are not suggesting that our writer deliberately misrepresented these statistics, but we are concerned that students do not always understand numbers, charts, and words like *majority*. We ask them not only to look up contextual information but to make sense of it when they do. Use these stories to help start conversations on accuracy.

Are the Name, Affiliation, and Organization Correct?

Entering a name or organization into a search engine is the simplest way to check spelling. If it is wrong, an alternate spelling will be suggested. Affiliation can be a bit trickier. In a story about high fructose corn syrup, an expert was identified as a "biochemist from Columbia University." However, while her degree came from Columbia, her job was running a consulting service to big food corporations. That mistake didn't mean that her comments were wrong, but it did color her position as a credible source. We caught that error by reading the person's web page.

Another student's submission had the following as part of an attribution: "Nilesh Parekh, article writer and a graduate of the University of Pune, in Indiana…" A check of University of Pune discovered it was located in India, not Indiana.

Is the Problem or Angle Right?

In a story on the devastation of acid rain, the author reported that levels of acid rain–producing sulfur dioxide in the air had dropped significantly over the past

few decades. Is this really a current problem? A quick check of Wikipedia found that this was a big environmental issue in the 1970s, but clean air regulations and new technology had reversed the trend. With nothing new to report, we decided to kill this story.

Colony collapse disorder, which is devastating honey bee populations nationwide, is another popular topic. However, most stories we received just recap national news, providing no new insights. The story on bees we did publish on *SciJourner.org* was about Missouri *not* being badly affected by the disorder. The author chose the right local angle to tackle the topic.

Is the Attribution Right?

Too often we get an attribution that looks like this: "according to *upmc.com*." Two concerns arise: first, is this a real site and second, is it credible? We have stumbled across URLs that don't work or that point to a site that makes no sense given the information.

This is a "small" problem, and easy to fix. In this case, *upmc* is the University of Pittsburgh Medical Center, a credible source. The author's job was then to include additional information in the attribution such as the name of the organization.

Another story offered the attribution to a "Dr. Park Nicollet," but when we checked the source Park Nicollet came up as the name of a hospital. Again, it is a case of a student not paying attention to details. We always kick these problems back to the author to fix.

What's Missing?

Probably the hardest item to fact-check is what is *not* in the story. A story on green schools pointed out that the author's school was not very eco-friendly, but an important voice was missing in the article. Editor's comments encouraged the author to check in with the principal about his position on the topic, which the author did!

In a story about underage drinking the author cited a state senator for increasing fines as a way to get at the problem. However, when we checked this fact, we discovered that it came from an article entitled "Experts: Higher Fines Alone Won't Curb Underage Drinking." There was no mention of these other proposals to halt underage drinking in the student's story. The author had picked the facts she liked, but ignored the substance of the discussion.

From our perspective, certain errors are more troubling than others. For an article to be published, we strive for complete accuracy, of course. Yet we don't always achieve it. Part of the beauty of publishing an article posted on the web is that it is subject to fact-checking by the entire world. Talk about a distributed community!

Context and Fact-Checking in Action:
Ideas for the Classroom
Practice Fact-Checking

A teacher began class this way: "Did you hear they're raising the price of lunch in the cafeteria?" As an excited (and outraged) discussion broke out, he said, "Isn't anyone going to ask me how much?" The students obliged. The teacher laughed and said, "That's my point. When someone says they're going to raise the price of lunch, the question you want to ask is 'how much!' Today we're going to look at an article that's missing all kinds of specific details, things like how much the price of lunch is going up. As I read it out loud, I want you to raise your hands when you hear places where you want more specific information." After clarifying that the school was not, in fact, raising the price of lunch in the cafeteria, the teacher went on to have a successful lesson.

At the end of this chapter, in Figure 9.2, we've included early drafts of student articles with real errors in them. These pieces can be used as read-alouds or for students to examine in small groups. Access to the internet will be necessary. Students should be encouraged to look for things that should be checked—and then determine how and where to check them. The lesson we call "Something Seems Wrong," available on *Teach4SciJourn.org*, does just this. Or teachers can create their own worksheets that include a variety of facts, some right and some wrong, and see if students can find all of the inaccuracies. Alternatively, students could submit inaccurate statements that the teacher might compile for distribution. Can they find one another's inaccuracies?

Here's another idea that can be used to help students help one another. By creating an ongoing Problems/Solutions chart, students can name and get advice regarding issues with which they are having trouble. On *Teach4SciJourn.org* we have begun an elaborated version of such a list, but in truth, it is better to do the naming of problems and the generation of solutions in class *with* students (see Figure 9.1).

Figure 9.1

A Teacher-Written Problems/ Solutions Chart

PROBLEM:	SOLUTIONS?
What to believe?	· Are all sources equally credible? · You can include different viewpoints · Are all sources recent/ up to date?
I can't put info into my own words	· Try saying it to a friend or tape recorder · Try drawing & then talking about it · Ask someone else & take notes

What to Do When Students Are Dead Wrong!

"I am interested in writing a news story," a chemistry student announced to his class during a topic pitching discussion, "about pregnant women and Red Bull. I heard someplace that if a woman is pregnant and she drinks Red Bull, her baby will begin to grow things on its back like wings. If she keeps drinking it, they will turn into wings."

The other students in the class were less than convinced about this topic.

"So you're saying that the baby will grow wings?" one asked.

"And the baby is born with a set of wings if the mother keeps drinking the Red Bull?" asked another.

"Yes," replied the young man. "I want to write about it." He was obviously serious and found this topic compelling.

(laughter among the skeptical students)

"It's true," he insisted. "I saw it someplace."

At this point the teacher stepped in and asked where the student had heard about this phenomenon. He said he wasn't sure but would find it again and show everyone. It was clear that no one other than the young man pitching the story saw any possibility of this topic meeting the accuracy test. He promised to come back with credible sources from the internet.

One of the visiting adults in the room jumped into the discussion.

"What about doing a story on the effects of caffeine upon a pregnant woman and the baby? You could examine the caffeine in Red Bull and any ill effects it might have on the pregnancy."

"Like growing wings and stuff," the student replied.

"Not that so much that," the visitor replied, "but something more realistic."

The student was not swayed. The idea of the wings was too compelling to be brushed aside by skeptical classmates or visiting adults.

The issue of strongly held misconceptions among students has been a challenge for educators. How do we best dispel ideas that are misguided and lacking credible supporting evidence? Clearly, it has been shown that merely correcting students or redirecting their thoughts will not address the underlying issues. There was no possibility of the young man leaving the classroom that day believing anything other than his idea about babies and wings and Red Bull. Yet, allowing misconceptions to fester and to grow will often interfere with real learning.

Several teacher responses were possible that day. One, permission could have been denied by the classroom teacher to pursue a topic that we know would clearly lead to a dead end. In the interest of time and efficiency, she could have responded as most editors of science news publications might and dismiss the pitched topic as science fiction. Or, she could have (like the visitor) redirected the student to a similar topic. Think of something more realistic and convince the student to pursue it instead.

Had the teacher insisted on either of these tacks the student may well have secretly harbored this inaccurate notion. Instead, the young man was

encouraged by the teacher (and many of his incredulous fellow students) to research the topic and bring back what he found for further examination.

A week later the student quietly told his teacher that he had decided to change his topic to something completely unrelated. He said that as hard as he tried, he could find nothing to support his original idea. He had actually come to doubt the notion of babies growing wings as a result of Red Bull.

Science journalism and its demands for multiple, credible sources can be a powerful tool for the classroom teacher in addressing the misguided ideas of students. But of greater importance, the students who work through the processes of science journalism come to decide for themselves what is convincing and what is not. Their ownership of concepts and beliefs, based upon a strong understanding of evidence gathered from reliable sources, fosters a seeking and acceptance of evidence-based thinking.

This is a great story to remember at a pivotal moment, when, for instance, students are looking for a topic or when they assert information that you know is wrong. Knowing when to sit back and when to interfere is part of the teachers' art.

A Final Word: When a Mistake Gets Past You

Please remember that perfection is impossible and newspapers publish corrections and online magazines fix errors all the time. This isn't always reassuring to a classroom teacher. Teachers don't publish papers. Instead, we assign grades, and when we mess up we can't issue a retraction. There is, of course, the possibility of changing the grade and asking the student to redo the work, but this isn't always possible. How do we deal with it?

The innocent mistakes are the easiest to come to terms with. A student who made an honest typo that gets caught down the line may have missed a chance to learn about proofreading, but science literacy goals are still in evidence. However, when students plagiarize or intentionally include false information, it's hard not to be angry. A student who makes up facts or expert interview answers is being deliberately deceptive—and not learning much about science literacy either. To head off plagiarism, it may be useful to identify problems and possible solutions ahead of time. Our ethical statement, SLAP, in Chapter 4, is one such effort.

But remember, the basic goal is to depend more on students and less on you, the teacher, when it comes to fact-checking. You will not always be there to help them out. If they are to become scientifically literate, they must hone their own instincts for sorting out inaccurate information (see Figure 9.2, p. 130).

Here are a few bits of advice, adapted from the practices of professional journalists, designed to do help students strive for accuracy:

1. If journalists don't know if the science/technology/medical information is right, they often look for others to talk to about the subject and ask their new sources to explain the science in simpler terms. Brainstorm with the student or class about who the article's author

---

Figure 9.2

Paragraphs to Fact-Check

- Black holes: we can make them now. The device that managed to do such a feat is known as the Large Hadron Collider, or the LHC for short. It may be dangerous, but it it's been stable and so far under control. And if everything goes according to plan, the LHC could help solve the mysteries of matter.

- A lot of murder and suicide cases are related to the listener digesting music with subliminal messages and gruesome lyrics. There's an actual story of a man who murdered his whole family after listening to a known rapper whose lyrics talk of killing his girlfriend, murdering people, and being possessed by demonic forces. Being careful about what we listen to is a wise thing to do; for the sake of our minds and thoughts. Music has gone a long way from having the upmost respect to now being used for shock value and high album sales.

- Aspartame is made by the weak bond of three chemicals; methanol, phenylalanine, and aspartic acid. Furthermore, the byproducts break down into formaldehyde, formic acid and aspartylphenylalanine diketopiperazine. It is possible to break down aspartame in long storage, in liquids, and in heat excess of 85 degrees Fahrenheit.

- "Gliese 581 is said to be around 20.5 light years away", says astronomer Margaret Turnbull of the Space Telescope Science Institute in Baltimore. That would take the standard space shuttle traveling about 17,500 mph over 70,000 years to get there and another 70,000 plus to get back. Space shuttles today use a controlled nuclear explosion to propel the shuttle forward. Once you're out of fuel you are out and there would have to be a resupply shipment out to Gliese.

---

could talk to. Another approach is to read about a topic on multiple sites, building up an understanding. Get the author to identify what sites might provide very general information and which are more technical.

2. Students worry rightfully if they "said it right." What words are to be avoided and which used? Look carefully at the language used by reporters as they write about the same or a similar topic. How does the language differ from the textbook language? Try finding a technical summary of the research and compare it to a press release. How does the journalist or press officer articulate the complex ideas, statistics or conclusions? Look for qualifiers in the article and recognize their importance. Look for the metaphors and comparisons they used.

3. Is the name, affiliation, and organization right? Is the website right? Encourage students to work in pairs or small groups as they go to a website or search the information. Have them talk about what they find by putting it into their own words. Does everyone agree that the attribution is correct?

4. Is the problem or angle right? Students should be encouraged to discuss what is interesting about the angle the author has chosen to take. Is there something unique about the region, family, situation, etc., that would change the story line in an interesting way? Did the author find the right places to research the topic? Did the writer talk to the right experts? Did these sources view the same issues as problematic?

5. What's missing? What questions do the teacher or other readers still have? What problems might the story's author note? In some classes, students read their stories out loud and ask for comments.

Remember, the work involved in situating stories within a context and making sure that the information provided is accurate and up-to-date is what teachers do most everyday in classrooms. You have practice doing these things; sharing the strategies you as a teacher use to provide information and fact-check it can be useful to your students. Such discussions also provide a venue for you to share and reflect on how you developed those instincts and where your own background information originated. Think and search with them. Working side-by-side with students, rooting around together on the information highway and trying to decide what is and is not a worthy stopping place, is a gift to the students with whom you work. And it's also lots of fun for the teacher.

Reference

Boyd, S. K. May 2, 2011. People with Down syndrome can do what you do! *www. SciJourner.org*

CHAPTER 10

GOING THE WRITE WAY

Certainly a class can learn a lot by just engaging in activities described in the previous pages and not actually writing an article. We believe, however, that writing moves the student to a deeper level of understanding and at the same time offers the teacher an important glimpse into student thinking. In this chapter we first lay out some organizational guidelines that support a writing classroom and at the same time help move young people forward in their science literacy skills. This is followed by a river of ideas designed to take students through the process of prewriting, drafting, and revision.

Setting Up a Writing Classroom

Before beginning a writing project, it helps to think through the goals of that assignment. These goals determine how much time you allot to different phases of the writing process, especially to revision. They may change from assignment to assignment or even from student to student. Here are a few assignment ideas we have seen work in SciJourn classes:

- *Students need to complete a prewriting assignment.* Students are required to do one of several activities, such as create a PowerPoint presentation or give a short talk about the article they would write if there was time.

- *A complete article with revisions is required for some.* Choose a set of criteria that students must meet in order to complete the writing assignment. Lessons are then designed to make sure students understand those criteria and offer them practice so they can meet the standards the teacher sets. One SciJourn teacher decided that her students needed to understand the importance of including multiple, credible perspectives in their articles. Anyone who turned in an article with suspect sources or from only one point of view (even if that point of view was represented by different sources) was required to revise. Those whose articles met the standard were encouraged to revise for publication, but this was not a requirement.

- *A single revision is required from all.* In these classrooms, teachers spend time on each of the SciJourn standards and encourage students to attempt a publishable article. All students receive feedback based on where they need the most help and are required to revise one time.

- *Students are required to write one draft, but must keep track of "lessons learned."* In this implementation, improvement on a "next article" is the real goal. In courses where more than one article will be assigned, the first article is often viewed as a test run for practicing SciJourn standards and journalistic form. After receiving feedback, students and teachers make note of what they did well and what still needs work (this could be classwide—e.g., everyone had trouble with attribution—or student-specific). These areas are addressed in a second article later in the semester.

- *It's not done until it's published.* One of our teachers expected all of her students to continue working on their articles until they were accepted for publication on *SciJourner.org*. (You may choose the school paper, another publication outlet, or create your own online site for this same purpose.) Her students, generally weaker writers, were used to "waiting out" writing assignments, i.e., wait for the final draft due date and throw something together the night before. This teacher wasn't kidding when she set the bar high. Each time students turned in their drafts, they expected to be finished. Each time, they got more feedback and more issues to attend to. At a certain point, they became so invested in their papers that they found it almost impossible to *stop* revising. Students in this class were published at an amazingly high rate despite the fact that they attended an "underperforming" school.

The point here is that your assignment—whatever strategy you choose—sets priorities and helps you organize your calendar and figure out when and where you have room for SciJourn.

The Writing Classroom: Organized Chaos

How do you keep a classroom where students are working on different topics and at different rates from feeling like a chaotic, disorganized mess? No matter what we suggest, a writing classroom is going to get a little chaotic (sorry!). Here are some ways to manage it:

- Make a consistent place for students to work and keep track of their work, such as a physical file folder, an *Excel* table, or online space like *Diigo.com* (see Chapter 7 or *Teach4SciJourn.org* for more on Diigo);

- Use Diigo's educator account to keep up with student research;

- Set up an article status board, such as the one in Figure 10.1 (p. 136);

- *Dropbox.com* provides free, online storage for files that can be both shared among a group and kept private;

- Keep "office hours" (to be held either during class or a study hall period) during which students individually quickly update you on their status.

Final thought: As all good teachers already know, you will get better at this as you learn what works in your school, for your students, and with your teaching style. This project does involve risk. Many of our teachers described themselves in the beginning as being far "outside of their comfort zone," but within a year or two, they saw SciJourn as an essential component of what they do *and* what they are good at. Keep track of your successes and failures for next time, talk to your colleagues in the English or journalism departments at your school, communicate with other SciJourn teachers online, make your students co-researchers in the teaching process, and be open to changing your plans when things are not working well.

Learning to Be Flexible

It turns out that writers come in different flavors. What works for one writer doesn't necessarily work for the next. What works for the teacher as a writer may not work for her student writers. Some writers need to rid themselves of the blank page at all costs. Getting *something* down is important in order to move on. Other writers will do almost all of their prewriting in their heads while they go for a run or shower or before they fall asleep at night. They simply cannot write until they are ready. What they finally put down, though, is nearly a final draft. Unfortunately, you don't know which students are which, and young writers don't always know who they are as well. In English classes, we often ask students to try lots of approaches and keep track of what works, but science teachers, who may assign only one or two articles, really do not have enough time to play around with different ways of doing things. To top it off, writing science news is different from

Figure 10.1

A Teacher's Guide to the Article Status Board

Throughout the process of article development some teachers use a status board posted in the classroom to track student progress. The board includes the name of each student and the steps for article writing. It provides a means to quickly observe the step at which each student is currently working.

The *Status Board* displays the following steps toward article completion:

1. Searching for a topic

2. Topic selected/ready to pitch

3. Research started/adding sites to *diigo.com*

4. Writing first draft

5. First draft submitted for editing

6. Finding photograph

7. Revising story based upon editor comments

8. Story submitted for publication

9. Searching for next topic

Pushpins can be used to track progress—green for moving along smoothly; red if the student is stuck at a particular level and is in need of assistance. Students move their own pins as they progress across the chart. Student progress is noticed at a glance.

Status of the Class

STUDENT	TOPIC	1	2	3	4	5	6	7	8	9

what students are used to writing. We've seen some interesting cases in which the students whom teachers consider their "best writers"—often because of their skill with the traditional school report—struggle with a news article precisely because it is so different. Finally, in the "real world" of journalism, there are deadlines. You can't just wait for inspiration to strike. It is the teacher's decision on how closely to follow the deadline mandate.

To make things even more complicated, journalism has its own set of stumbling blocks that can delay students. Some students may be waiting a few days for an interview; others may begin with a promising topic only to find out the research is too difficult or too new (see Chapter 6: "What's Your Angle?"); those doing surveys may be waiting for a certain number of responses or for help calculating the statistics (see Chapter 8: Original Reporting: Interviews and Surveys).

The bottom line is this: All students will not move through a writing assignment at the same pace. Unless you want to purposefully delay the students who are ahead, which can result in frustration and decreased motivation, you'll have to learn to love or at least tolerate a classroom that feels more like a workshop of elves than soldiers on the march. But think about it—if you told Ernest Hemingway, Willa Cather, and William Faulkner to write a short story, would you expect them to proceed in the same way? Due dates make sense, but you need to build in flexibility.

There are, however, significant advantages to being flexible. You can set a window for collecting rough drafts rather than an absolute deadline, which means fewer papers to respond to at any given moment. Students who are hung up on their research can get some help from the librarian rather than being rushed along for the sake of the calendar. Those who finish early can engage in some of the following activities:

- *Encourage students to add more to their stories.* This may be an original graphic or web-borrowed and attributed photography, captions for the graphics, or another interview or a survey. They could also convert part or all of the article to a podcast or short movie.

- *Employ students who finished their articles as "experts" in the classroom.* Depending on the skills of each student, some could help with fact-checking, others with checking the credibility of sources. One teacher found that the students in her class who had published articles on *SciJourner.org* were suddenly the celebrities of the class. Other students were eager for their input and these students were able to ask the kinds of questions the teacher and the editor had asked of them (e.g., "What's your source for this?" "Why is this source credible?").

- *Allow students to move to hands-on activities supported by their article.* Share the example of the student who did a piece on how increased CO_2 levels made students sleepy and then began measuring the CO_2 levels in her class at various points in the day.

- *Invite students to begin a second article.*

- *Respond in the comments section to online science articles, building from the SciJourn standards they have enacted in their stories.*

A final note on pacing: We know that there is a limit to how flexible a teacher can be. At some point, you do have to move on. For students who are straggling behind out of laziness or apathy, a firm deadline will be necessary. Likewise, students who are waiting for someone to return a call or e-mail may have to go on without that interview. This happens in real newsrooms, too. We have all heard news broadcasters say things like "could not be reached for comment" or "did not return repeated requests for an interview." A student can add this to their article, too.

Ideas for Prewriting

Writing is a process that minimally includes prewriting, writing, and revision. Writing in this sense begins long before one confronts the dreaded blank page. All of the activities described in this book so far can fruitfully be thought of as prewriting activities: finding an angle, identifying sources, developing interviews and surveys, collecting background information—all of it. But now it is time to put pen to paper or fingers to keyboard. Is there anything you, the teacher, can do to help?

Writing teachers like to point out that the writing process is recursive—we go back and forth among prewriting, drafting, and revision during the process. Some activities we call "prewriting" may actually take place after a student has begun to write, such as when a student suddenly realizes she's missing a key piece of information and heads back to the internet. The recursive nature of writing is part of what makes writing *about* writing so difficult—it's hard to write in a linear fashion about a nonlinear process. While the page demands that the ideas be put down in the final article in a logical order, remember that students can be encouraged to skip some steps or go back and return to others later in the process.

Prewriting loosely includes everything students do to prepare to write. Students often go through "stages" of prewriting: Understanding the assignment, brainstorming for a topic, researching for information and planning to write. As students sit down to write their first drafts, the teacher hopes that they (1) understand what a successful article looks like; (2) have a workable topic and angle; (3) have found multiple, credible sources; and (4) understand the research they are about to describe.

The Importance of Mentor Texts

If you've been following our advice, your read-alouds have focused on content or our science literacy standards and not on writing. However, along the way writing will probably pop up as a point of discussion. After reading a particularly compelling lede, for example, you or the students may comment on it. At the time, you may not even use the terminology *lede* or point out *why* the lede is compelling, but the idea that beginnings should arouse interest will be planted.

Once your students are ready to begin writing, it is useful—perhaps essential—that you revisit some of these articles or continue sharing read-aloud/think-alouds (RATAs). If an article got a particularly positive reaction, you can look at it again with the class to figure out what about it was so good. Because the students are already familiar with the content, it will be easier to focus on form. Ask them, for instance, which of the many articles you read do they remember. Why did they remember this one? Or show them three articles and discuss which was the best and why.

As students are being asked to write their own articles, they are much more likely to attend to your read-alouds because they have a *need to know*. They will be looking at old articles with new eyes and trying to find solutions to problems they are encountering as writers. During the writing process, you may begin one class period by just reading and discussing several ledes; another class period can include reading context paragraphs from several different articles. Bottom line: do not stop reading aloud just because students are now in the writing phase.

As you work through this project the first few times, it might help to begin collecting examples that can be used in various ways. An article doesn't have to do everything well to be a useful model. One might have a great lede, another might provide a good example of how to attribute different sources in different ways. One might show how an author made numbers memorable or digestible, another might provide a model for quoting an interviewed source. Identifying articles for their writing features ("good narrative lede," "good attribution of a nonprofit," etc.) and keeping them on hand somewhere in the classroom will provide a resource for students who are interested in writing for publication. Also, check *Teach4SciJourn.org* for useful examples.

Weak examples from the internet are also powerful reminders of what *not* to do. Use them. Students have no personal investment, no one's feelings will be hurt and you might even discuss how these weak articles could be made better.

Although not specifically a science literacy skill, learning from model texts is something that will serve students well in the future. As Cathy Fleischer and Sarah Andrew-Vaughan point out in their book *Writing Outside Your Comfort Zone*, when competent adults are faced with an unfamiliar task—writing a resume, for example—the first thing they do is seek out examples to imitate (2009). Our students should learn about the importance of models and value the models they view as particularly successful. They also learn a lot from noticing how incompetence displays itself.

Do They Understand Their Research?

Many students need time to think, talk, and write about their research before they can write. Pushing them to get a draft done without allowing for this often leads to plagiarism—they have to put something down and they lack the understanding needed to paraphrase. Giving students time to talk to one another (and to you) can help. Short written assignments can also precede the rough draft. You may want to build in one or more of the following strategies, which are all also excellent formative assessments:

- *Stable groups for the duration of the project.* These groups meet together often to discuss how the work of prewriting and writing is going. They can also meet online, in study halls, or after school. For students who benefit from structure, these groups can meet regularly; use guiding questions to focus their discussions.

  You might, for example, designate who will report to the group on any given day. Groups of five work well for this reason, since Mondays can be John's turn, Tuesday's Kayla's turn, etc. The reporting person is expected to (1) update group members on their progress, (2) identify problems they are encountering, and (3) ask for suggestions.

- *"Speed" feedback.* This can be set up like the "speed pitch" (Chapter 6), with desks arranged in two circles, an inner and an outer, facing one another. The outside person describes his research and his partner asks questions and gives suggestions. At the end of a few minutes, the inside person describes her research. For round two, the inside people move one space clockwise and the process is repeated.

- *PowerPoint.* The power pitch (Chapter 6) can serve well as a form of feedback and used again to update classmates, even showing a single slide as ideas and information falls into place.

- *Diigo dialogues.* Check in on each student's diigo site. You may ask students to write a note or tag their bookmarked sites; there is also a place for you to write back.

- *Outlines and webs.* Some students benefit from creating a formal outline; others like to create webs of connections (which can be done by hand or using a computer program). Because these are so short, you can respond quickly to a class set.

- *Student–teacher conferences.* Checking in with students doesn't have to take a long time and you don't have to talk to every student in a single class period. Keep track of who you've talked to on a clipboard or in a notebook. Just asking students to tell you what's new can give you a sense of what they understand and where they're confused.

Of course, there are many other methods of getting students to talk about their research (Figure 10.2). The point, however, is that most students need some thinking and talking time before they jump into writing, especially if they are writing in a new genre.

Figure 10.2

Elevator Speeches

It can be helpful to have students talk about their research before writing. This gives them an opportunity to organize their thoughts, make sense of what they've learned, and get feedback from their audience. An elevator speech—a quick overview deliverable in the length of an elevator ride—can be a time-efficient way for students to accomplish these goals:

1. Prior to the class meeting, all students should have chosen topics and completed some research. Ideally, they would be close to writing first drafts.

2. Explain what an elevator speech is, or simply tell the students that they will have 30 seconds to 2 minutes to summarize their research for their classmates. Ask each student to quickly take notes or draft a short speech about their articles. Since students have already been researching, you do not have to give them a lot of time (alternately, they could have prepared their speeches for homework prior to the class).

3. Divide the class into two groups. Half should scatter around the room. These students are in the "elevators," with their speeches/notes in hand. The other half should form groups of two or three.

4. Groups should be instructed to move around the classroom, from classmate to classmate, listening to the speeches. At the end of the speech, they should ask clarifying questions and offer feedback. Once they have done so, they should move to another speaker. Because the moving students are in small groups, there should always be an open speaker to find. Even if the students are not asking excellent questions of one another, the speaker will benefit from articulating his or her thoughts several times. If some groups remain at a particular speaker for a lengthy amount of time, be sure to remind them to move on.

5. After a set amount of time, students switch roles.

6. When the time is up, students should be given a short amount of time to write down some notes based on the questions they heard or to debrief on the experience as a class.

Next Steps

Depending on the student's experience during the elevator speeches, it may be time to do additional research, refine the topic, or begin drafting the article.

Teaching Journalistic Form

Although we do not suggest spending weeks and weeks on journalistic form, at some point students need to know the form of a science news article if they are expected to write one. Some teachers wait to discuss form until after students have already written something very rough, while others talk about it earlier. Ideally, students will have noticed lots about form from the RATAs you've been doing. The goal, then, is to make what they have already noticed explicit. The ideas that follow are ways to do just that.

Spend Time in Class Comparing the Traditional School Report With a News Article

When she was a high school English teacher using SciJourn, Angela and her students spent time looking at the news article and the report. Together, they came up with a list of differences, which Angela compiled into a chart. The list included the structure (beginning-middle-end versus inverted triangle), citation style (MLA format versus in-text attribution), and kinds of sources (library/database versus credible websites). Now that she looks back on this list, Angela realizes that it is much too focused on superficial differences. It misses the essential differences in content between the two genres.

When working directly with young people, we have often heard our professional editor ask students what they know about writing. After eliciting comments about the importance of a thesis statement and conclusion, he turns to them with a comment like, "That's great. Now put everything you know about that kind of writing (or the five-paragraph essay) and put it into a box, tie it up tight, and put it away for use in another class."

For students who have had a lot of instruction in report writing, more exploration of the differences and similarities between the two genres may be beneficial. Comparing and contrasting is a powerful teaching technique. However, science news writing is not simply about mastering a new formula. We believe the differences between a successful essay and a successful science news article are deeper than this.

From a science literacy perspective—and remember, that's what we are ultimately about, not writing—there is real value in the way the news article is structured. First, it insists that the main science idea is forefronted and that means that students need to know or recognize the main idea. Context is also essential; as we say elsewhere, students need to understand if the science presented is new or well established. If it's well established, references to it need not be explained in great detail, but if it's new or not fully explored, such as the health risks of tattoos or the release of elk into the Missouri countryside, the article's author is bound to check what different stakeholders might say about the information, even if they are quite convinced of their own position.

Handouts, Leading Questions, Graphic Organizers

One of Angela's writing students once asked her to "teach it like math—show me the steps." Writing *isn't* math (and math, when taught well, isn't really even math in the way the student was implying, but that's another story), and Angela was reluctant to create "steps" that would make writing seem like it was. However, this student needed something to get her started. She couldn't comply with Angela's advice to just "get something down." She wasn't that kind of writer, at least not yet.

Getting something down on paper seems like an easy task, but for some students it takes a lot of confidence and maturity to move forward. After much practice at a given writing genre, students might be able to proceed on their own, assured that they will know how to fix things later. But in the beginning, they may need something more. For students who struggle to get words on paper, hearing that you can always start over later isn't reassuring, it's terrifying. For those students, something less formal—a handout with a list of questions, a graphic organizer to fill in—might help them start writing things down without the pressure to make it look like an "article" just yet.

The following documents might be helpful for students or teachers looking to scaffold writing:

- The inverted triangle handout (Figure 2.2, Chapter 2)

- Characterizing a news article (Chapter 2)

- Student version of the standards (Chapter 4)

- The SAFI or CPR assessment tools which can be turned into an assessment guides of sorts (explained at the end of this chapter)

Please exempt ready writers from required exercises, because participating in them takes time away from other, deeper activities related to their composing process and may, in fact, eliminate the pleasure of "figuring it out" for themselves. On the other hand, even the best writers need models if they are going to figure it out. In general, we suggest teachers and students together move from whole (a familiarity with the genre), to part (looking at specific elements of the story using the standards), to whole (constructing and reconstructing a story based on what they now know).

The Dangers and Benefits of Focusing With "Exercises"

When we first began using the SciJourn project, we thought it was necessary to teach students a lot about journalistic form before they were allowed to begin writing. To do so, we created many "exercises." They spent time on ledes, reading ledes, critiquing ledes, and writing their own ledes. They learned about the inverted triangle structure. Using what they learned, they examined a cut-up

article and tried to put it back together. They had to underline the *who, what, where, when,* and *why.* At the end of the year, we asked them to evaluate the experience. Here's what we heard: "I wish we had written earlier." Or, "You spent all this time talking about it—I just wanted to do it."

In our eagerness to prepare students fully, we had delayed the process too long. As a result, many lost their motivation. Some of their topics—the possibility of hacking electronic voting machines being used for an upcoming election, for example—had gone from being timely to being old. We spent too much time preparing for writing and not enough time writing.

Now we strongly encourage our teachers to get students writing earlier, even if the students don't yet know about journalistic form. In her second year using SciJourn, one of our teachers summed up her thoughts this way: "Last year, I was regimented. I had all these lessons about journalism and I was going to use them. I went through them one after another after another. This year, I'm more flexible. The kids need to write. Just get it down. We can fix it up later." Undoubtedly, the most common sentiment expressed by teachers at the end of their first year is, "I wish I had started them writing earlier."

Our educator tendency is to scaffold and support, to release gradually. But our SciJourn experience, demonstrated time and again, is that these "supportive delays" may in fact make authentic engagement difficult if not impossible. Such support can rob students of the sense that they own the process. We need to make room for them to fall off the SciJourn balance beam; falling and getting up is part of learning, we believe. To belabor this metaphor for a moment longer: What makes learning to walk that balance beam in the presence of a teacher different than falling on one's own is that the teacher can note the point at which you fall, suggest some specific modifications, and encourage you to get back on.

A case in point: Rose Davidson, whose students were so familiar with the five-paragraph essay that they could chant the parts of the format in unison, felt a strong need to present exercises so that her students clearly understood journalistic form. Interestingly, despite her emphasis on differences between a news article and the five-paragraph essay, many of her students still wrote five-paragraph essays for their first drafts. Was that bad? Not necessarily. A five-paragraph report that meets all the criteria can either be transformed with the help of an editor into a more journalistic style in the revision process or left as is. It is all a question of "what is good enough." As an aside, many of Rose's students eventually published their stories, to her own and their delight. But they did not, they could not, learn about the changes their stories would need in advance. They needed to get something down on paper and then go back to it.

Drafting

How do you know when you're ready to write? It's a question we asked teachers who had just written an article as part of our professional development. For many of them, the answer was "you don't." The topics they had chosen to write about were personally interesting and meaningful; most said they could have gone on researching indefinitely. Others dreaded the writing because they were scared to try a new genre or because they simply didn't like to write at all. We can almost guarantee you'll have students who fit into all of these categories in your classroom.

Where to Begin: Not Necessarily at the Beginning

Like the 12th-grade students Janet Emig studied in the 1960s and 1970s, our student writers agonize over the first sentences of their articles, often refusing to write at all until they've thought of the "perfect" opening (1971). Those first lines usually stay in the paper through the final draft, no matter what happens to the rest of the story. (This isn't a problem only with students; journalists do it, too.)

Find ways to encourage students to jump in without waiting for that ideal opening. Tell them directly, "Get started. Don't waste time perfecting the lede." Teachers have successfully used various strategies to make this happen. Our colleague Nancy Singer used to sit at the computer and type the first few sentences for students who were really stuck, just to get rid of the blank page. An added advantage was that students found it easier to change their opening lines when they revised—they had no attachment to her words!

Other teachers simply tell students to begin with the second paragraph. From the get go, make it clear to students that they will need to write more than one draft if they want their article published. The first draft is only a starting point. The teacher or the peer editor uses it to respond to big issues such as a lack of credible attribution. Some students may find it easier to write if they know their first drafts will get completion points or only written feedback, not an evaluative grade. Sharing the experiences of other writers is also helpful.

If students have been using a bookmarking tool like *diigo. com*, much of the work of the first draft may already be done. Students can pull the material they've highlighted together into a document and then work on paraphrasing and connecting information logically.

"I sat down and wrote even though I knew I had more work to do. I took the plunge."—Hannah Johnson

"I was paralyzed by the fear of not being able to come up with a good enough lede for my article. I finally began my article with about three possible ledes and I left them all at the top of my paper with the intention of choosing the best one later. This strategy allowed me to begin the writing process easily." —Pamela Hughes-Watson

"As I write, I'm able to scan my Diigo account and pull quotations directly from my profile page. For further information or source identification I can click through to the website. I found this technique to be extremely efficient. I'll definitely teach this trick to students."—Andrew Goodin

Don't be alarmed—in fact, feel encouraged—if first drafts end up looking like a bunch of notes on a page, rather than an article. After years of plugging ideas into an essay formula, they may have forgotten how to be storytellers. Not so long ago, when they were children, teens understood story structure and how ideas build one from the other. Remember, and remind students, a good *SciJourner.org* article (or movie, podcast, infographic, or photo caption) basically tells a story: an up-to-date, accurate, contextualized and relevant story based on multiple, credible sources, but a story nevertheless.

Preparing Students for Next Steps With the Edit-Aloud

Writers, professional and fledgling, often have difficulty writing if they don't know how their work will be evaluated. While a teacher can provide assignment sheets, rubrics, and other methods for describing how they will assess writing, most students do not understand what success looks like without seeing and discussing successful (and unsuccessful) models. Working with mentor texts—described earlier in this chapter—will be helpful, but the edit aloud is another highly successful strategy. We have found these think-aloud strategies actually more helpful than grading rubrics.

Grading rubrics—though designed to make explicit criteria for success, offer little sense of *why* the teacher or editor cares about the issues he or she has defined as important. The edit-aloud gives students a chance to see inside the head of their teacher, which in turn helps them when they write. An edit-aloud is just what it sounds like—the teacher verbally responding to a paper in front of the class. Put a sample paper up on the overhead or distribute copies to the class and then read the paper out loud, voicing your thoughts as you do so (Figure 10.3).

Again, there are different approaches to the edit-aloud. One method would be to simply note and then ignore any problems with the article that you consider unimportant at this stage. For example, if you do not plan to mark grammar and spelling mistakes, say ahead of time that you will purposefully ignore these issues and not comment on obvious errors as you edit out loud. Alternatively, note them verbally, but do not mark them on page. If students jump on these errors, you can re-explain that, although you value proper form, you are only going to grade these drafts on X, Y, and Z. This is a way to clearly reinforce your purposes and help students prioritize what matters in their own writing.

Figure 10.3

An Annotated Edit-Aloud

***Directions for edit-aloud*:**
First, read the article to the class (project the unedited version). Ask for ideas about how an editor might respond. Then, read the article aloud again as you take on the role of editor—use the editor comments as prompts for thinking aloud.

Better Bring the Bug spray

Due to the new disease called White nose syndrome that is spreading in bats the mosquitoes, beetles, and moth population will be increasing. The first white nose syndrome case has been confirmed in Pike county.

White nose syndrome, which originally began in New York , is a disease that causes bats to grow a white fungus on their wings and noses. They also begin to wake up during their hibernation period and search for food; the bats lose all of their stored fat and energy. Bats will either starve to death or freeze to death.

Bats are very misunderstood creatures. They are our first line of defense against insects. In a 2006 study shown by the international bat conservation department over 68% of people think bats are useless and nasty creatures. "Over 1 million bats that have been killed were estimated to consume over 700,000 tons of insects in one year," said the Missouri Conservation Department. That equals the weight of about 175, 000 elephants. This epidemic is not only bad for the community but it is bad for farmers. Farmers will begin to see more beetles and other insects that can harm crops around. If you see any dead bats around caves do not handle them, the affects of the white nose syndrome on humans is unknown, contact your local conservation department and tell them about your find.

Comment [A1]: Are we really sure that those populations *will* increase? It's best to say they *could* increase because we don't yet know.

Comment [A2]: Pike County where? According to whom?

Comment [A3]: When?

Comment [A4]: Many readers might not understand how bats could lose all of their fat and energy stores if they are searching for food. You need to explain why.

Comment [A5]: According to whom?

Comment [A6]: Check the name of this – do you mean Bat Conservation International?

Comment [A7]: This should read Missouri Department of Conservation.

Comment [A8]: Nice use of an analogy to help readers understand this large number. Did you read this somewhere? If so, and to avoid plagiarism, be sure to include an attribution.

Comment [A9]: What is the source for this?

Comment [A10]: Additional editor comments: You did a good job explaining why this disease is important even though most people rarely see bats and don't like them. The headline *Better Bring the Bug Spray* is catchy, but it really doesn't say what the story is about. If a reader sees only the headline, is it obvious that the story is about bats?

Another tactic is to comment on a large number of issues as you think aloud, but then prioritize them when you are finished reading. Say, for instance, "In my comments to the authors I would probably write: 'Do the following three things in your next draft,'" and then specifically list what you want them to do. Students will be able to see how you decide which issues are most pressing. Eventually, you can get the group or even individual students to do an edit-aloud. An activity like this also prepares a student for revision.

Whichever approach you take, finish the edit-aloud by creating some concrete feedback and post it so that students can see and remember what you expect. If you are going to grade the articles with a letter grade, explain what grade the sample paper would get.

An edit-aloud is an especially effective activity to do a day or two before you plan to collect rough drafts. Do it too early, and students probably won't remember what you emphasized; if you wait until after you've collected rough drafts, they won't be able to immediately apply what they've learned.

An Instrumental Interlude

In this chapter we have offered a variety of strategies teachers can use to support student writers. We call this section "An Instrumental Interlude" because it contains two instruments we have developed to (1) help students write better, (2) help teachers respond more effectively and concretely to student work, and (3) move students toward publication (the subject of Chapter 11). These instruments can effectively be used in the classroom at various points: during read-alouds, as you introduce the writing assignment, as peer response tools, and to evaluate student work. Loosely, we call them "assessment tools"—tools to assess professional articles, sample articles, and student articles.

The challenge of assessment stems from the reality that science journalism articles do not fit into neat, formulaic packages. Although our standards have served as guidelines in determining article quality, there are no hard-set rules that lead to clearly measurable outcomes. Quality articles come in different shapes and sizes. For example, although we know the importance of multiple, credible sources of information, one might ask how many sources are required to make an article credible? We cannot say that six sources are better than five, but we do know that one is not enough.

Moreover, our goal is to help students write quality articles based on an internalized understanding of what quality articles look like, not based upon checklists or rubrics where points yield a blind grade. And it would be inappropriate for a student to get a decent grade while leaving out an essential element like multiple sources in an article that scans well according to a scoring guide or rubric.

Yet teachers and their students want guidelines to help them stay focused on the important issues when responding to student work. We have found two "systems" that have proved helpful: Calibrated Peer Review (CPR) and the Science Article Filtering Instrument (SAFI). What follows is a description of each and ideas for how they can work separately and together.

Calibrated Peer Review, the Other CPR

We have always been believers of peer review, but were having trouble making the process work. Students tended to focus on mechanical corrections or offer bad advice; in our project they were not attending to issues articulated in the standards document. Moreover, it was hard for teachers to imagine their students being able to offer good feedback to one another. Yet, to keep our publication going, we had to find a way to reduce the editing load on our editor and teachers. We also wanted to find ways to help teachers so that they would be comfortable assigning writing even more often.

According to an *Educause* article (ELI 2005), the Calibrated Peer Review (CPR) system is an effective solution because it combines two processes that scientists find familiar: calibration and peer review from unidentified reviewers. A CPR assignment follows clear steps, as follows:

1. Students receive a writing assignment, craft their response, and submit it to the website.

2. They then "calibrate" themselves as reviewers by reading sample essays (high, medium, and low) and responding to instructor-created questions. If their responses do not match the instructor's answers closely enough, they must recalibrate.

3. Students respond to the essays of their classmates.

4. Students complete a self-assessment.

At the end of an assignment, students receive feedback that shows how closely their own assessments matched those of their peers.

The idea behind CPR intrigued us, even though the program is most widely used in college classes. We were concerned that some of our schools lacked the needed internet access and that our standards were difficult to turn into the kinds of questions this system required. However, the idea of calibration was fascinating, and we brought it to teachers at our next professional development meeting. Their responses mirrored ours. They weren't sold on the entire program, but they loved the idea of calibration essays, particularly if we could provide annotated, scored versions of edited articles. Being good teachers, they all had different ideas of how they might use such essays in their classes.

Although we had always encouraged teachers to provide students with models and had even suggested using "imperfect" articles for read-alouds, to date we had not provided authentic student articles, filled with the real kinds of mistakes students make. This was what the teachers seemed most excited about.

However, we confronted a few hurdles in developing a databank of calibrated sample articles for the CPR system. As previously mentioned, one

problem was creating questions that emphasized our standards, but were also able to be answered the same way by trained reviewers. Questions we developed worked to some extent, and included:

1. Topic: Does the topic of this article meet at least three of these criteria: local, narrow, focused, relevant, timely, interesting angle?

2. Sources: Does the article contain multiple sources of information with various kinds of expertise and/or perspectives?

3. Sources: Are there places where information should be attributed and places where it is not needed?

4. Sources: Is enough information provided about the sources of information to establish credibility?

5. Sources: Are any sources included that are of questionable credibility?

6. Context: Does the article provide details about how the topic fits into the wider world, such as how much it costs, how many people are affected, how it compares with other technologies, whether it has political/economic consequences, or whether or not the story content is accepted or controversial?

7. Relevance: Does the article answer readers' questions?

8. Accuracy: Does the author explain all of the science accurately? Are all unfamiliar terms defined?

9. Rank this article (high/medium/low)

Any updates to our version are available on the SciJourn website, *Teach4Sc-Journ.org*.

A second challenge was identifying medium-quality essays. In the first batch we created, some teachers weren't sure which article was low and which was medium. It may be that the most likely reason an article falls into the low pile is topic. If the topic or angle is poor, the article is doomed until the topic is refined—no matter how well the article may score on any of the other standards. Finally, we decided that perhaps the issue wasn't medium versus poor, but rather what could CPR and teachers do to enable students to improve their articles. In this sense, CPR helps students focus on what is going well in their stories.

A Science Article Filtering Instrument (SAFI)

The *Science Article Filtering Instrument* (Figure 10.4) is a homegrown effort to offer assessment guidance developed by Laura Pearce, SciJourn researcher, author, and award-winning teacher. It is not a point-based system, but rather enables the user (teacher or peer-editor) to filter and assess quality based on specific attributes, indicated by checking a particular shape.

Figure 10.4

Science Article Filtering Instrument

Author _____   Topic _____   Date _____

To assess article quality, check one shape beside each characteristic. Use the key to determine level of quality.

Does the article being assessed contain any: (if unknown, check ◇)

1. Stereotyping?   Yes △   No ◇
2. Lies?   Yes △   No ◇
3. Advertising?   Yes △   No ◇
4. Plagiarism?   Yes △   No ◇

NOTE: Reject article if one or more △ checked.

Key
* If one or more △ are checked, the article is immediately rejected.
* If one or more ○ are checked, the article is in need of significant revisions.
* If one or more □ are checked, assessment depends upon ratio of ◇ and ○ checked.
* If every ◇ is checked (and no △, ○ are checked), the article is ready to be submitted for consideration as a published article.

The article being assessed contains:

(check one)

	Complete or nearly complete	Needs improving	Absent
5. Two or more sources of information that are credible and properly attributed.	◇	□	○
6. Viewpoints from more than one perspective, when appropriate.	◇	□	○
7. A clear explanation of the science content, which indicates a basic understanding by the author.	◇	□	○
8. Assertions that are reasonable; if not attributed, are within the general knowledge of the audience.	◇	□	○
9. Information that is relevant to readers.	◇	□	○
10. Information that is factually accurate.	◇	□	○

Notes to author:

Date _____   Evaluator _____

For example, if the article is exceptionally well written yet only one triangle is checked in Figure 10.4, the article is rejected. The triangles represent violations of our SLAP ethics (Chapter 4), which are extreme infractions—or as Laura calls them, capital offenses. Nothing elsewhere in the article can make up for one of these major missteps.

On the other hand, if every diamond is checked, the journalist has included each of the key elements. These are the articles approaching publication quality. If any one of the diamonds is not checked, revision is likely needed in that targeted area.

"The SAFI was really helpful in keeping me anchored. It kept me from focusing on grammar when content was more important."
—Beth Grasser

Using CPR and SAFI

Our teachers have given us several ideas for using these feedback tools, both separately and together, with student writing and with sample articles. What follows are some of the most popular ideas. We encourage teachers to offer additional ideas and adaptations to the comments section of the website. Once again, sample articles, along with our editor's answers to the CPR questions, are available online.

- Have students "calibrate" with sample articles several times. Teachers often begin by working through the CPR questions with a sample article as whole class. Students next get a chance to work in small groups; those who are ready can try to do articles by themselves. This may take place over several weeks; some students may need several tries before achieving an acceptable level of calibration. Be sure to allow enough time for discussion of student answers. This can be done even in classrooms where a written article is not assigned.

- Focus on calibrating to only one or two of the standards with the sample articles. See if students can match your (or the editor's) answers only on sources, for example.

- Use the SAFI on student rough drafts as a peer-editing tool. If students can recognize the major problems with each other's articles before you receive them, they've gained an editing skill and you've reduced your paper load.

- Use the SAFI for providing teacher feedback on early drafts; save CPR as a feedback tool later in the process. You might want to think of the two tools as actual filters. The SAFI is wide mesh; the CPR is finer. What gets through SAFI will not necessarily be high-quality according to CPR.

- Use either CPR or SAFI with students in small groups to open up conversations among students. This can be done with sample articles

or the students' own writing. Just like teachers benefit from talking to one another about student work, students benefit from hearing the opinions of their classmates. Depending on your class, you may want to have a discussion about proper feedback etiquette if students are using their own writing.

References

Educause Learning Initiative (ELI). 2005. Calibrated peer review: A writing and critical-thinking instructional tool. *http://net.educause.edu/ir/library/pdf/ELI5002.pdf*

Emig, J. 1971. *The composing processes of twelfth graders.* Urbana, IL: National Council of Teachers of English.

Fleischer, C., and S. Andrew-Vaughan. 2009. *Writing outside your comfort zone: Helping students navigate unfamiliar genres.* Portsmouth, NH: Heinemann.

CHAPTER 11

IT'S ALL ABOUT REVISING:
MOVING TOWARD PUBLICATION

I f you are one of those teachers whose goal is to help students get published, please understand that you are making a major intellectual and time commitment. We can almost guarantee that no student will get a first draft published; in fact, since an article (and one's science literacy skills) can *always* be improved, we see revision as a fundamental purpose of SciJourn writing.

In days of old, prior to important research on Writing Across the Curriculum, students were only expected to revise papers in their Language Arts class—if at all. Writing assignments required in other classes were typically short and almost always one-shot assignments. Turn in the paper, receive the grade, and move on. We now realize, however, that this approach missed an essential learning opportunity. Learning is, in fact, revision, and good teachers see the value in having students correct their mistakes on homework and retake failed tests.

Revised writing provides a similar avenue for students to get a second (and third and fourth) chance at understanding and further developing content and skill.

In classrooms, we have seen excellent teachers working with motivated students who develop and demonstrate critical thinking as they listen to read-alouds and critique professional science stories. They pitch great story ideas, read research, and interview experts. Yet, time and again, these same students turn in articles that lack qualities we view as essential to science literacy. This isn't surprising, although sometimes we are still surprised by how universal the problem is. Students may grow in their ability to consume science news critically, while still struggling to produce articles themselves. Although we believe a classroom that focuses only on consumption will improve student science literacy, producing an article demonstrates a deeper level of understanding. Try your best to head off certain issues. In our experience, students struggle with:

- *Attribution.* Students may be able to critique sources in another person's story, but until they recognize the importance of backing up their own words with credible, named sources, they are not demonstrating the level of science literacy we seek.

- *Context.* Students have trouble connecting their topics to the wider world. They do not include details like the number of patients suffering from a disease, the cost of a current technology, or the research that their article is building upon. Adults who are scientifically literate see these connections and ask these kinds of context questions.

- *A sense of what's important versus what is a detail.* Students seem to have trouble forefronting important information. Learning to identify what is important about a story is a science literacy skill.

Although we believe revision is important, we also know that it can be time-consuming and challenging to incorporate into a curriculum. Teachers must prioritize by requiring their students to revise for what they deem important. Not every student will revise to the point of "publication." Although it is up to the teacher to determine the level (or number) of revisions required, as our editor tells us repeatedly, tenacity is what finally gets articles published on *SciJourner.org*. Many published pieces have gone through two or three revisions, some as many as eight or nine drafts.

The teachers' discomfort with revision may only be outdistanced by that of their students. Here are some process activities that might help them:

1. *Find some distance.* Trying to revise a piece when you are too close to it is almost impossible. Allow at least a day or two between the first and next drafts. The more distance, the easier it is to make changes.

2. *Performance helps.* Whether you read your story aloud to yourself, your teachers, or your peers, the very act of reading helps create that

distance mentioned in point 1. For teachers trying to figure out how to break the news that the article is really problematic, ask the student to read it to you and we can almost guarantee that your admonitions will no longer be needed.

3. *Marking teacher/editor comments you do not understand is useful.* Decide if you agree with the reviewer and if not, ask for more information.

4. *After completing an article, summarize it in a single sentence.* Look to see if that sentence is reflected in the nutgraf.

5. *Cut your paper into paragraphs.* Ask a friend to help you organize by what is most important, next-most important, etc.

6. *Spread out revision over time.* If a teacher expects multiple revisions, make sure that you understand that getting published or getting a great grade may involve multiple revisions.

When to Start Over

One of our teachers asked her students to write two stories during the course of a yearlong class. During the first semester, they wrote their first stories and received feedback on them. When they began their second stories second semester, they were given the option of returning to their first stories and revising or starting again. While some chose to start over, others stuck with their original work. Both groups moved through a second in-class unit on science journalism and turned in their articles at the end of the unit.

For teachers who choose this approach—a second unit on science news with the option to revise or start over—even students who stay with their original topics may want to change their articles significantly. One way to encourage students to think about their topics anew is to read each other's articles and then write one additional question they have about the topic on the back of the article. Each student ends the activity with several ideas for a new but related story. This suggestion is actually about responding, the subject of the next section.

Responding to Student Work Individually: Different Kinds of Feedback

Just as student writers come in various flavors, so do teacher responders. Some like to offer targeted tips for one student, others focus on whole-class discussion. Some have a conference as the center of their feedback.

For students who've turned in a very rough draft—whether it is the kind that looks like bullet points or sentence fragments or something with narrative flow that makes little sense—or seem to be stuck on how to move forward on their story, the following standard practices may help them turn the draft or research "pieces" into a coherent story. Try offering suggestions such as these on an "as needed" basis. Said differently, these comments are designed to help the teacher in responding to student work, but it is not a list to be broadcast or distributed to the entire class.

Play to the Personal

If you ask a teen why he or she is interested in a topic, especially a health topic, you most often find that the issue touches them or someone in their life. Rather than the generic "Diabetes is a disease that…" opening, we ask for the personal connection. Many of our favorite *SciJourner* stories fall into this category. Often we find these personal connections almost tacked on at the end of a story. For example, in an article on lung cancer that mimics an encyclopedia entry, the narrative ends with a statement about the student's grandfather trying to quit smoking. We encourage writers to move these personal connections to the forefront and structure the story around them.

To do that, we typically push the teen to interview the affected person or to speak from their own experience. If the person with the malady can't talk or has died, someone who witnessed the events is a good choice.

The other approach to this type of story is an imagined scenario: "What if you woke up one morning with a tumor growing out of your leg?" It is almost always a bad choice, and probably inaccurate. Also, avoid cutesy or humorous approaches. Given that people die from sickle cell anemia, we strongly suggested to a student that he not lead with the idea that this mutation is different from the one that created the comic book hero Spiderman, for instance.

Compare those bad ideas for ledes with: "I was only the eighth person diagnosed at that time," my aunt, Julie Foster, tells *SciJourner*. "I was relieved to have a diagnosis and sad that there was no cure" (Lovell 2010).

If an author is the main character, she might begin with something like, "I woke up in the middle of the night with my head pounding and throbbing" (Ragland 2010).

In the case of an illness, the lede should describe the initial symptoms or events that first signaled the disease or a major event as the disease evolved— all in the words of the person or family affected. It sets the stage for a dramatic story.

Intersperse comments on the progression of the disease, from symptoms to treatment, with the science. Thus, the opening describing a personal event in the disease might be followed with the more straightforward reporting on the numbers of people with the disease, a formal list of symptoms, and a scientific explanation of what is happening.

End the story with the status of the person now. "Even though I'm not fully cured yet, I have the confidence that I will heal completely. I am positive that one day, when I look in the mirror, I will see the man that I saw in the picture from not so long ago" (Pasupuleti 2010). It is this unfolding personal story that encourages the reader to finish the article.

As noted earlier, we don't shy away from difficult topics; it is our experience that individuals and families want to tell their story. It is also important to recognize ethical issues as they may arise. We always warn authors to make sure that their subjects are comfortable with their story being told to the world. If the story involves a minor, it is probably good to get permission from the parent or guardian. In some cases, you may want to use a false name. And make sure that the topic conforms to school policies.

Surprise the Reader

Lead with something unexpected and build on that throughout the story. "Compact Fluorescent Light bulbs, known as CFLs, are famous for saving energy and lowering electric bills. However, they can harm the environment and humans. Pretty ironic?" (Carlson 2011).

"Did you ever imagine that life could start at the very depths of the ocean floor? Some scientists hypothesize, with help from studies on "black smokers," that underwater sea vents could be originators of life" (Kelly 2010).

Offbeat topics are another form of surprise: "Have you ever purchased fresh produce only to discover that it has rotted a few days later? If so, you know how aggravating this experience can be. Infomercials claim to have the solution: Green Bags" (Haas 2011).

The surprising story builds from the lede. For example, the CFL story immediately moves into an explanation of the mercury and how much of the metal is in these light bulbs, how many of these bulbs are in use, and what can be done to minimize their harmful effects. The ending looks forward to energy-efficient lights without mercury.

State a Problem

"When you wake up in the morning you take it for granted that you will be able to get out of bed on your own. But for someone who suffers from a physical disability, even the act of getting out of bed is a team effort" (Goebel 2010).

"Less than 50% of youth in foster care in the United States graduate high school, according to statistics from the Orphan Foundation of America" (Redus 2009).

The details of why can follow. Here, too, is a place for the personal. In the foster care story, the author interviewed one teen who went on to college and one who dropped out of high school.

Have Appropriate Fun With the "Small" Story

When the disease is not serious, the problem is not big, and there is no surprise the writer can have some fun.

"What would you do with $10 million dollars? How about a new car? Just design the car of your dreams and enter it into the Progress Insurance Automotive X-prize 100 mpg Car Challenge" (Johnson and Turner 2010).

"Have you ever suffered from stinging hand pains or numbness in your hand from typing on a cell phone? If you have, you are not alone. Around 25% of the students at Francis Howell High School in St. Charles, MO are right there with you" (Bond 2011). In this story, the author described the problem and its treatment in medical terms, but ended the story with "but we all know the only real cure for it is to put the phone down," in essence, using a lighter touch to bookend the story.

Let's Talk: Responding Through Conferences

We mentioned conferencing with students as a prewriting activity. Conferences are also an excellent way to respond to drafts and move students toward publication. The conference is an opportunity to listen as well as talk. Written responses may address the wrong issues or misread the student's level of understanding. Conferences allow for more clarity and, when compared to the time it takes to read and respond in writing, can actually be more efficient.

Conferences need not be formal nor must they take place at the teacher's desk. Conferences can be part of what you do as you move around the room, checking in on student progress. In real journalism offices, reporters update their editor on progress all the time.

To get the maximum benefit out of conferencing, you may want to try some of the following. These strategies work well even if you are working with students who are not particularly interested in publishing their work.

- *Don't try to conference with all students during a single class period.* Keep track of students, topics, progress, and conferences on a clipboard or in a notebook. Be sure to make a check mark or include a quick note when you conference with a student, so you know who you've talked to and who still needs attention. Some students will naturally want to conference at every possible opportunity, while others will happily hide in the corner.

- *Ask the student to lead the conference by identifying one or two concerns.* This is a challenge for students who may ask you to look at "everything" or "edit it for me." (SAFI, discussed in the previous chapter, can be helpful here.) The only way students get better at recognizing their own strengths and weaknesses is to practice. Encourage students to ask for help on the big picture issues you've identified as important in your lessons. If a student asks for help with a concern that you think is secondary to something else, listen to why he or she is worried about the issue and explain what you think is more pressing and why.

- *Keep the conferences short and focused.* Once the primary concern(s) with the article have been identified, do not prolong the conference by pointing out additional flaws or areas for improvement. This will only make the conference too long. It will also give the student too much to do in one revision.

- *It is not always necessary to read the entire article during the conference.* Depending on the student's main concern, you may just look at sources of information or at the first half of an article. Do not feel obligated to always read and provide feedback on every word.

- *Be sure the student takes notes during the conference.* If you take the paper from the student and do all of the writing during a conference, you take on all the responsibility and work. In some cases, this may be no different than collecting the papers and returning them with your remarks. Take advantage of the live nature of the conference by asking the student to do the writing. This gives you the advantage of seeing what the student writes down, and it forces the student to ask clarifying questions that might not get asked if you do the note taking.

- *Be clear that you are not covering everything in one conference.* Some students naturally see a conference as a "contract" between you and them. If they do everything you mention in the conference, they assume the article will automatically get an A. Discourage this kind of thinking by ending the conference with a clear indication that addressing these concerns will move them forward in the writing process—but will not be the end of that process (e.g., "I look forward to seeing the next draft," or "Once you've done X, we should talk about Y").

- *Conference with groups of students who have similar concerns.* For example, if you know several students are struggling to put their stories into context, pull them all together for a small group conference. Together, look at the first article and model the kind of feedback you would give the author. For the remaining articles, ask the whole group to contribute to the conference.

Additional Thoughts About Conferencing

During this project, we have thought a lot about the hybrid world we are creating in classrooms. This world is one that draws heavily on the real world of science journalism, while adapting its practices for use in the classroom. In real newsrooms, an editor would not conference with a journalist in the same way that classroom teachers conference with students. But this classroom practice, when used in conjunction with science news writing, can be a powerful tool that supports both better writing and science understanding.

This hybridity became very clear when a group of researchers from the project visited a science teacher's classroom to demonstrate how to conduct a writing conference. Student papers had been sent to the researchers via e-mail prior to the conference and the researchers had provided written feedback. This written feedback was fairly short and, in some cases, a little snarky. The researchers were not impressed with the papers at all.

However, their impression changed dramatically during the writing conferences. In talking with students about their work, fascinating stories emerged. For example, a student had written an article about video game addiction after a friend actually received treatment for this problem. Another was interested in the physics of ballet, because her ballet teacher was also a trained scientist. None of this information appeared in the original articles, which were generally superficial five-paragraph reports. Hearing these stories helped the researchers guide the students toward their next drafts. The advice given in the conferences was, in some cases, very different from the advice they provided in writing.

For the researchers, who did not know the students or their topics prior to the visit, these conferences were somewhat time-consuming (around 10–15 minutes each), but the SciJourn team learned so much! Teachers could do the same thing these researchers did, but much more quickly. Teachers *know* their students and their classes. Teachers know where most of the student topics came from (or they can easily ask)—they can easily see if the interesting anecdote told during a warm-up pitch session (Chapter 6) was somehow left out of the rough draft. Teachers may have watched their students searching—they know where certain information came from or have further details about what appeared to be poor attributions. Information that took several questions for our researchers to elicit took much less time for the classroom teacher to gather. A conference exists in order to make the most of the personal relationship students and teachers already have.

If you need ideas about what to say during conferences, see the SciJourn Editing Hierarchy (Figure 11.1)

Responding to the Whole Class

When a large number of students are making the same kinds of mistakes, good teachers respond to the entire class (or a subset) at once. These responses can take the form of handouts, PowerPoint slides, or mini-exercises. One technique is to do a more targeted version of the edit-aloud to demonstrate specific problems often found in articles. Display only a few sentences or paragraphs of text and see if students can spot the problems (see Figure 11.2, p. 165).

Responding in Writing

Early in our professional development sessions, we asked teacher participants to respond to two sample student papers. This pretest was designed to capture the teachers' initial editing tendencies. When we analyzed the way teachers responded to student papers, we found that about half of their comments were related to the form of the article, not the content. In particular, teachers focused on grammatical errors and typos, or issues like whether or not the capitalization

Figure 11.1

SciJourner Editing Hierarchy

What does it take to be published in *SciJourner?* Although editing is not an exact process, it is possible to state some consistent goals. Overall, *SciJourner* is targeted at getting writers to meet SciJourn standards. To achieve these targets, teachers can best help their students by focusing on content editing over copyediting (spelling and grammar).

Below are some guidelines to help you both edit your students' stories and talk with them about how they can improve their writing.

Articles that will not be considered for print include those that have:

- *Violations of the SLAP ethics.* Be particularly vigilant for cut-and-paste sections from the internet. [Sample conferencing comment: A reporter would get fired for doing this—and with good reason.]

- *Flaky or bad science, attribution to bad sources.* [Conferencing comment: I don't believe this story. _____ is not an appropriate source for this story because_____.]

- *No attribution at all.* [Conferencing comment: Why should I believe any of this without attribution? Your readers need to know where you got your information.]

- *Science details are factually wrong.* [Conferencing comment: I don't believe that you understand the science behind _____. You should check your sources or consult _____.]

- *Not up-to-date.* [Conferencing comment: This is old science/data/technology. How can you find newer information?]

- *No science in the story.* [Conferencing comment: How can you get some science in this story?]

Stories that need major revisions include those that have:

- *Insufficient attribution.* [Conferencing comments: Says who? According to whom? Where did you get this information?]

- *Missing or misspelled details such as researcher names, institutions, dates, etc.* [Conferencing comment: Point out specific instances to student and ask them to check all of facts, names, institutions, etc.]

- *Missing background information, such as how many people are affected by the disease, why would anyone do this, etc.* [Conferencing comments: Point out specific items that need more information.]

- *A single, but credible source.* [Conferencing comment: This is a single source story. What other sources could verify this information?]

- *A small factual error but the rest is fine.* [Conferencing comment: Point out the specific factual error and ask the student to double-check the other facts in the story.]

- *Attribution does not seem to correlate to a source, such as a statistic that isn't on the cited website.* [Conferencing comment: I can't find this fact/number/comment on this website. Please check it and all of your other citations.]

- *Direct or indirect quotes that are not attributed.* [Conferencing comment: Where did you get this quote?]

- *Jargon that is not understandable to a high school student.* [Conferencing comment: Please "translate"; How would you explain this to your friends?]

Continued

Figure 11.1. *SciJourner* Editing Hierarchy *(Continued)*

Minor revisions for students or you as the editor to fix include:
(When making changes, it is important to respect the author's voice. That is, editing should not make the story "yours" instead of the student's.)

- *Articles written as a report, but everything else is fine.* [Rewrite and eliminate the concluding paragraph.]

- *Poor lede.* [Show student example of a good lede, suggest a new one lede or let it go]

- *Badly or overwritten.* May include too many "you" or "I think," wordy, extra paragraphs, etc. [Work with student one-on-one to tighten the writing; remove editorial comments]

- *Little experimental detail but the rest of the science is OK.* [Let it go or ask for more detail]

was appropriate. When they did react to content, their attention was drawn to the factual correctness of information; they queried the dates of a flu pandemic, for instance. Very rarely did pre-SciJourn teachers comment about credibility or sources, context, or relevance—the very items that rank high for our editing (Figure 11.1).

After learning about our science literacy standards, these same educators taking our posttest changed their focus, attending less to grammar and form and more fully to issues of content. Although they made fewer comments, their comments were designed to promote science literacy and, we would argue, were more meaningful.

Pointing out surface errors is relatively easy and not very helpful. Resist the temptation. Yes, we do understand that that's what teachers may have done on your papers. For those of us who eventually came to consider ourselves good writers, it is natural to conclude that this sort of feedback helped. Even those who grew to hate writing may find the mechanical errors easier to identify (or may think they are easier "fixes" for students) than what we think of as global concerns. What's more, teachers often feel a sense of obligation to mark these kinds of errors, since we know that students will one day be held accountable for correct form, and we may even fear that someone will think that *we* do not know the difference if we let such errors stand.

In the first year of the project, our professional science editor—who had very little experience with high school students—was discouraged by some of the writing he encountered. Seeking to be helpful, he ended up writing more in his responses than some student authors had in their first drafts. As he was asked to deal with greater numbers of submissions (like a real teacher!), he realized first, that this approach was unsustainable and second, that it did little to develop science literacy (which we note, yet again, is our goal). Instead, he now focuses in highly specific ways to address content, attending to issues like grammar and mechanics only when papers were close to publication. In fact, he

Figure 11.2

Targeted Edit-Alouds

The excerpts below can be used to point out specific problems student authors run into when writing their articles. The comment boxes include the actual feedback provided by our editor. Additional examples are available online.

Many of the articles excerpted below were eventually revised and published on SciJourner.org. You and your students may want to look at the changes the authors made in response to editorial feedback. Search the story topics on the website.

Who cares? Topic isn't made local or relevant:
These are titles from first drafts of student articles. Ask students to think about the topics of these articles and brainstorm ways to make them more local, narrow, focused, and/or relevant to a teenage audience.

Titles:
Is ozone good or bad for people?
Caffeine Addiction and Caffeine Supplements
Luminol
An Aspirin a Day Might Keep the Doctor Away

> **Comment [a1]:** You need to narrow the topic and really talk about some issues. For example, if tropospheric ozone is high in St. Louis, then what is the issue? Even better, if it is high in winter or summer, then what is the issue?

> **Comment [a2]:** You are circling a couple of interesting topics, but you don't really state what you are covering. Is the topic caffeine supplements? Or are you writing about caffeine addiction? You need to keep your topic narrow.

> **Comment [a3]:** What is the story here? This reads like a textbook entry. What is new? What is the issue? Does luminol always work? Why would anyone read this story?

> **Comment [a4]:** Although interesting, this is pretty much taken from a press release. What have you added that is new or different? Why should I read this version, when I can find the press release on dozens of sites on the Internet?

Who else could you talk to? Missing perspectives:
Read the excerpt below and ask students who else might be interviewed on this issue.

Student Article Excerpt:
  Many people at our school share very different opinions on whether our school is doing a good job or not. Most of the student body believes the school could be doing a much better job like how Zach Shultz, a student at Francis Howell High explains, "The school is not doing enough to create a more eco friendly facility and the last thing that on the school board's mind is the efficiency rating of the school, the school can do much more".

> **Comment [a5]:** If you are feeling brave, I would suggest you talk to the Principal or Vice-Principal about the question of how eco-friendly the school is. Tell them that the students think you can do more and get their response. Yes, it sounds scary, but they know about SciJourner and probably would be more than willing to talk to you. Really!

Who Says? Missing attribution:
Ask the students to raise their hands every time you get to an assertion that should be attributed. See if students notice where this student author begins editorializing.

Student Article Excerpt:
  But what exactly is an alternative medicine? They are "practices used instead of standard treatments that generally aren't recognized by the medical community as standard or conventional medical approaches," according to the National Cancer Institute. So, some examples of these would be yoga, eating healthy, herbal remedies, and etc.
  However, they are now more commonly known as complementary and alternative medicine. This term means the use of a combination of alternative medicines and complementary medicines which are conventional medicines (common everyday medicines that are found in your medicine cabinet such as Tylenol, Advil, and etc.).
  It is believed that these medicines are becoming more popular because scientists have finally come to their senses that nature's remedies are the best. After all, most alternative medicines have fewer side effects than conventional medicines.
  So where have these alternative medicines come from since they have been around for a long time?

> **Comment [a6]:** according to who? Where did you find this information?].

> **Comment [a7]:** This is a pretty strong statement. I think many scientists would disagree with this and say that they have been looking at nature all along for ideas. And are natural cures always the best? Yoga doesn't treat cancer as well as many pharmaceuticals.

> **Comment [a8]:** according to who? Again, this is a mixed picture.]

Continued

Figure 11.2. Targeted Edit-Alouds *(Continued)*

Alternative medicines were probably around ever since the first generations of humans, but the first documented use of alternative medicine was Yoga in 2698 B.C. This would conclude that alternative medicines are nearly 5000 years old!

> **Comment [a9]:** says who?

How does this connect to the wider world? Context is missing:
Read the opening paragraphs of this student article and ask students what kind of context information they would want to know.

Student Article Excerpt:

About a year and a half ago, my grandfather, Christopher Dalton age 56, made "the hardest decision of my life. I made the decision while I was working in my office, and a few of my friends asked me if I wanted to smoke with them outside. I looked out my window and saw that it was pouring down rain, and I said no, not right now; it is raining. Later, when I saw them outside in the rain smoking, I told myself there is no point to keep smoking." He decided to quit right away, "cold turkey," which made me wonder, "What happens to your body as you progress through your life after you quit "cold turkey"?

> **Comment [a10]:** You need a background/context paragraph here. It should answer such questions as How many Americans still smoke? How many try quitting each year? How many are successful after 1 year, 5 years, etc.? How hard is it for people that have smoked as long as your granddad to quit smoking? How many try to go cold turkey? Do most try an assisted approach, such as the patch or gum? All of these stats are somewhere out on the web, most likely at the American Cancer Society, but maybe also CDC or NIH or US Government Human and Health Services. You can also look at the US Surgeon General.]

What does that mean? Science that needs explaining:
Read this excerpt and see if students can pick out where more information or explanation is needed to make logical sense.

Student Article Excerpt:

By schools operating in a more environmentally conscious way, not only does it create a much cleaner environment and reduce pollution, but lowers the school's cost of operating and we can use that money for much more important things. For example, a research conducted by the U.S Department of Energy showed that one of the highest expenditures for a school is their energy cost. Public schools have been spending over 8 billion dollars a year in energy cost! It is a fact that schools can lower their energy by up to 20% if they manage it correctly. Schools can do such things like replace light bulbs with energy saving bulbs which itself can reduce its cost by 30%. Another great way is simply to turn lights off when a room is not in use or turn the computers off at night which accounts for about two third of a school's energy. The Best way would be to educate the school staff and students on energy consumption and ways to save energy.

> **Comment [a11]:** This next section is great stuff, but I'm concerned that you have not explained it all correctly.

> **Comment [a12]:** You have a logic problem. In the first sentence you said energy costs could be cut by UP TO 20%, but in the next sentence you say 30%. Do you mean that just the lighting bill can be cut by 30%. Not clear why there is a mismatch.

> **Comment [a13]:** This should say "energy usage." You also have a problem with the way the sentence is written, it implies that ONLY the computers account for 2/3 of energy usage—is that right? Where did you find this information?

Read this excerpt and see what science questions students have.

Student Article Excerpt:

In the 1990's Scientists and Researchers were looking for a way to prolong the lifespan of produce in preparation for a trip to Antarctica. They ended up discovering an area in Japan that had been successfully housing and storing produce in mountain caves for thousands of years. The mountains were continuously dark and dry and were composed of clay called oya which contains zeolite, a naturally occurring desiccant.

Before the invention of Green Bags, the April 1999 publication of *ChemMatters* says that pellets of zeolite were used by various produce companies to remove ethylene from the atmosphere of produce when it was being transported. Zeolite is saturated with Potassium Permanganate ($KMnO_4$), which is a good oxidizing agent. The Mosby's Medical Dictionary defines an oxidizing agent as a compound that readily gives up oxygen or accepts hydrogen or electrons from another compound. In chemical reactions an oxidizing agent acts as an acceptor of electrons.

> **Comment [a14]:** says who? Is that just your statement or did you find that in one of your sources?

> **Comment [a15]:** What do you mean by continuously dark, and what does that have to do with zeolite?

> **Comment [a16]:** What does the permanganate have to do with protecting fruit? Is it oxidizing ethylene? Is it in green bags?

made virtually the same number of content edits on submissions he viewed as publication-worthy as those he saw as unpromising.

It was from his content edits that we derived and checked the appropriateness of our standards. He and everyone working on this project believe that all students can improve their science literacy by engaging in content-level revision. In fact, the science news article, if properly understood, provides a scaffold that almost forces us and our students to attend to content.

A side note from the English teachers among us: correcting students' grammar doesn't get students to attend to mechanics either. If you care about mechanics, look for error patterns and get students to make corrections themselves; don't do it for them.

Editors and Teachers

The goals of a teacher and the goals of an editor are, of course, different. While the editor must think about journalistic style and publication, a teacher can be fully engaged in this project without ever commenting on the wording of a lede or grammatical mistakes. But what can teachers learn from editors that is classroom-useful? What *can* teachers reasonably expect of themselves and their students?

One teacher, looking at the comments our science editor provided, described herself as feeling like she'd "failed." She had already responded to the student articles and thought she did a pretty good job, yet our professional caught many more issues. She was not worried about the form edits that she missed. After all, she was not an English teacher, she said, and so she didn't expect to be able to edit for journalistic style. However, she was bothered by the fact that our editor questioned students' sources that she had let slide by.

This experience got us thinking. Our message that teachers don't need to worry about form had gotten through. However, we were concerned, just as she was, that some of the more critical problems were being missed. Our advice? Give yourselves a break! No professional editor starts off able to spot these things, and teachers cannot learn to spot every problem overnight. Our editor has spent years learning about credible sources and has recently gained more experience spotting students' bad sources.

Teachers also benefit from talking to other teachers about student writing. One of our teachers commented, "I think that the dialogue with colleagues is essential in this process. It's nice to be able to discuss questions and concerns as you encounter them." A databank of articles along with their editorial feedback is available online at *Teach4SciJourn.org*. Ideally, several teachers in a school or a district would be trying SciJourn at the same time and could use these articles to have discussions together.

The features of professional editorial feedback that teachers can immediately put into practice have to do with tone and specificity. All teachers who go through our professional development training write their own science news articles and receive professional editing as part of the process. They are often surprised at the kind of comments the editor provides; these comments are direct, specific, and usually not sugarcoated with the compliments teachers often include. While some teachers worry that students might be upset by this kind of editing, in our experience, students find it refreshing. One student pointed out that teachers are indirect, saying things like "you *could* do this, or you *might want* to do that," whereas the editor would say something like, "you need more information about the number of people who suffer from this disease."

In our research, we've also noticed this difference in tone. Teachers often ask general questions (e.g., "Can you explain this more fully?") or couch their comments with words like "maybe." The editor makes direct statements or asks for specific additional information. The experience of being edited helps teachers see what this kind of editing looks like and often inspires them to change their style. As one said, "I think that constructive criticism should not be equated with being mean. Our job as teachers is to teach students how to use their knowledge to operate in the real world, taking into account real-world expectations and consequences."

"Actionable" Written Feedback

Unless you have clear goals for your feedback, you may find yourself responding to student writing in a way that is time consuming and unhelpful to the larger goal of improved science literacy. The teacher's intentions may vary from assignment to assignment, or even from student to student, and all students do not need to be working on the same issues. An important point to keep in mind is the idea of concrete, specific feedback, something one of our teachers described as "actionable." The "action" a student should take may be different depending on whether or not you are planning multiple revisions, but specificity is always key.

What is important from a teacher's perspective is maintaining focus and keeping notes so you can remember what you sought from their rewrites or peer-edits. Our science editor constantly grapples with the question: "How good is good enough for publication?" Writing teachers face a similar question, and, while it may be difficult to come up with an answer, the question is worth considering before you spend hours on a set of papers: How good is good enough for an A or a C grade? This is one reason why we suggest setting up your priorities before you begin a writing unit; it's worth reminding yourself of them before you sit down with student papers.

Nuts and Bolts: What to Do With That Stack of Papers

No matter how much work you've put in before collecting student drafts, at some point you are going to have to sit down with the submitted articles and provide some feedback. This can be a daunting task, one many teachers put off for weeks. While all teachers and classrooms work a little differently, many teachers find the following steps helpful:

1. *Read through the entire article before you begin marking the paper.* This may seem like a waste of time, but it's only once you've read the

article to the end that you can provide proper feedback. You do not want to spend time commenting on a sentence near the beginning that turns out to be completely unimportant to article's main point. It also helps avoid line-by-line editing of articles with serious problems, such as plagiarized stories, missing or inaccurate science, no attribution, or unworkable topics.

2. *Limit yourself to two to three concerns per paper.* We've stated this advice over and over again.

3. *Be clear, specific, and concrete in your comments.* In our discussions with students about writing feedback they typically get, many mention the confusing marks they've received on their work. Teachers circle a word or underline a sentence without giving the author any indication of what these circles and lines mean. Others cross out entire lines of student writing without saying why.

   When you mark within the text of a student's paper (or in the margins), be sure to include enough information that the author understands the comment—and attach the comment clearly to the student's writing. For example, if a student wrote a sentence that included no attribution, draw an arrow to the sentence and write in the margin "Who says?" If the attribution is included but vague, you may write something like, "What is this person's expertise? Where does this person work? Why should I believe them?"

4. *Do not mark every in-text problem, even if it is one of your identified concerns.* If a student consistently has no attribution in the article, rather than mark every instance within the text, address the problem in your holistic comment at the end or beginning of the article.

5. *Include a useful final comment at the end of the article.* While in-text or margin comments are about specific sentences or paragraphs, the comments you include at the beginning or end of the article should be holistic. Sometimes, teachers confuse a holistic comment with one that is overly general. We have heard students complain about vague statements in their end-of-article comments—statements such as "include more details" with no indication of where or what kinds of details might help.

   Depending on the nature of the article, the holistic comment can be used to do one of the following:

   o Address problems that keep the article from working on even a basic level. No matter what you do in class to avoid major article problems, some students will probably turn in papers that are simply cut-and-paste snippets from the internet, that have topics that are more appropriate for a book than a 500-word story or that show a lack of basic understanding. The holistic comment can point out these large flaws and offer suggestions for moving forward.

- Provide suggestions that apply to the article overall, but not necessarily to a specific place in the current text. If the article could use information from another point of view, mention it in the holistic comment. Comments about a lack of contextual information or about the order of the information might also be provided at the end of the article.

- If the article needs extensive revisions, use a holistic comment, such as "can you find more credible medical sources than *ehow* or *about.com*?" to help the student prioritize what to do next. You may even choose to provide a numbered list, explaining the next steps you suggest. Remember to walk the fine line between being specific and not being so explicit that you have done the revision yourself (a tricky line to walk sometimes!).

- Holistic comments can also be used to provide an overview of the problems that the in-text comments illustrate. If the article lacks attribution, for example, you may say something like, "Remember that any information in your article that is not generally accepted to be true must be attributed. I've marked a few places where you need attribution, but you should review the entire article to catch them all."

6. *Be constructive and specific with your positive comments.* Many teachers choose to include at least one positive point whenever they respond to a student article. In (almost) all cases, this can be done—but be sure that your positive comments are just as clear and specific as your critiques. Students certainly benefit from an in-text comment like "good contextual detail" that is clearly tied to a piece of data about the number of people suffering from a disease, but they don't get much out of a line like, "Good job!"

7. *The best articles should be covered in red ink (or green or blue or purple).* It is ironic that bad papers get the lion's share of editing. In SciJourn, we reverse that. The closer a story is to being published, the more we edit it. Thus, the more red ink, the more we like the story! The numbered list also works well as a holistic comment for these almost-ready-to-be-published articles. In fact, we have found that if we list what needs to be fixed as 1, 2, 3, etc., students get the message that they need to fix them all.

8. *When assigning grades, it sometimes helps to wait until you have commented on all papers.* Many writing teachers read and comment on all papers without writing any grades on them. Instead, they

separate the papers into piles, identifying target papers. (This is a clear A, here's a clear B, etc.) Then papers in between are assessed: "Is this paper more like the As or more like the Bs?" If you prefer, avoid letter grades until later in the process and shift to words like *excellent* and *good* as the students develop as writers and researchers. This can be particularly useful if it is the first time you have ever tried an assignment or for the first 10–20 papers you grade.

When the "Stack of Papers" Is Virtual: Using Track Changes

Collecting hard copies of student drafts can be a waste of paper and an organizational nightmare. In response, many teachers now require students to submit papers electronically, either by e-mail or through an online dropbox. Rather than printing out all of these papers in order to respond, some teachers prefer using the Track Changes feature of Microsoft Word. (Some students seem to enjoy it, too. In one of our schools, students used Track Changes to do peer editing and found it so helpful they were soon begging to use it in other classes—and even teaching their teachers how.)

The advantage of Track Changes is that students can immediately see your edits and comments, which are marked in a vibrant color. It also handles comments from multiple editors by assigning the different individuals different colors. Sophisticated Track Changes users can even selectively jump from edit to edit, accepting or rejecting them with a single keystroke. A guide to this tool is provided on the book's supplementary material website.

Final Thoughts on Responding to Writing: The Evolution of an Editor

Over the course of this project, our professional science editor has adapted his editing style to meet his and the teachers' many goals. This is a project about science literacy; the primary goal of his editing is to improve students' access to and understanding of science content. Yet he is also responsible for putting out a science news magazine. His editing is designed to move at least some stories toward publication, although he does hope that some students learn to write tighter, clearer prose because of their SciJourn experience.

Our editor's newly developed philosophy of editing is to turn content edits—edits about issues that have to do with our standards—into questions that the student must answer. He may sometimes provide hints or additional factual information, but he rarely corrects content concerns himself. With writing problems, though, he does just the opposite. He will directly change a student's writing to make it simpler or smoother, but this happens only when he believes the students have understood the issues they are writing about. What we don't want is to have students press "accept all" in Track Changes—blindly

accepting whatever the editor has put on the paper—and feel they are done with an article.

Often our editor's changes are accompanied by explanations of the edit. In this way, he is hoping to teach students about writing while maintaining the focus on content. Figure 11.3 (pp. 173–176) illustrates this point.

Of course, our ultimate goal in responding to student work is to enable students to revise on their own, to change our teacher voice into a little voice within their own heads. It's about helping students recognize and respond to the issues described in this book on their own, when you are *not* around. That's what science literacy is finally about.

References

Bond, T. January 14, 2011. Blackberry thumb. *www.SciJourner.org*

Carlson, A. March 19, 2011. CFLs: Friend or foe to the environment? *www. SciJourner.org*

Goebel, H. April 23, 2010. Horses and healing. *www.SciJourner.org*

Haas, A. March 20, 2011. Do green bags really work? *www.SciJourner.org*

Johnson, D., and K. Turner. August 15, 2010. X-prize challenge. *www.SciJourner.org*

Kelly, K. June 7, 2010. Are black smokers architects of life? *www.SciJourner.org*

Lovell, B. May 26, 2010. Orphan diseases get noticed. *www.SciJourner.org*

Pasupuleti, T. December 12, 2010. Pemphigus: A disease that changed his life. *www.SciJourner.org*

Ragland, N. July 27, 2010. What's wrong with my head? *www.SciJourner.org*

Redus, D. July 12, 2009. "Exceptions" to the foster care system. *www.SciJourner.org*

Figure 11.3

Student Articles Edited Using Track Changes

The student's original writing is in standard type, all editorial comments and changes are in bold.

Example 1

This is an interesting topic, but you have not written a very persuasive article. Other than the study published in the Journal of Pediatrics [which is highly credible], you don't quote any first-hand credible sites. Aren't there websites maintained by hearing specialists?

Also, I can't publish the accusations against Apple or their earbuds without some sort of support from experts.

And why don't you do your own survey on how loud your fellow classmates play their ipods? That would strengthen the story a lot.

[We don't need to imagine anything. Your lede is the 12.5%, which is a real number.] ~~Imagine thirty years in the future and you're sitting in front of the television and the only entertainment you are getting from the program you used to love is reading the subtitles.~~ According to the Journal of Pediatrics, **[actually, it is not the Journal of Pediatrics, but rather the researchers who published the work in the journal. You need to say who they are, where they are located and when this was published.]** 12.5 percent of kids between the ages of 6 and 19 suffer from loss of hearing as a result of using ear phones turned to a high volume. ~~Now 12.5 percent of children does not seem to be that big of an impact on society to you but just think if you were~~

Continued

Figure 11.3. Student Articles Edited Using Track Changes *(Continued)*

~~in that percentage and things you were hearing seemed like murmurs~~. That 12.5 percent of children with hearing loss is from headphones, not including stereo or concert music [did they say that stereo or concert music was a risk?].

It seems today when you walk down the halls of a school you see almost everyone with headphones in their ears, rocking out. [you don't need to narrate the story, just tell the facts.] ~~Now, thirty years in the future will those kid's headphones turn into hearing aids?~~ A study from www.hearing-it.org [are you sure about this URL; it comes up empty for me.] found that most cases of hearing loss ~~in current times~~ are caused from excessively loud noises or environments and that by 2025 over 40 million people could be [this isn't definite] ~~will have been~~ diagnosed with some form of hearing loss. ~~In more simplistic terms that conclude that over~~ \or 13% ~~thirteen percent~~ of all Americans will have some form of hearing loss in the future.

IPods' are the most widely known mp3 player devices, but even then Apple has done nothing of public knowledge to try and force a limit on sound output of their devices [says who? What is your source?]. It is not only the iPod that is having effects on a person's hearing all over the world. The stock headphones that are included with the Apple devices seem to be very cheaply made and created for the "one size fits all", and rest on the outer part of the ear lobe allowing surrounding environmental sounds to penetrate more easily [what is the source of this accusation? Are you just saying things? Don't do that, we will get sued.]. Due to the fact that those ear-buds do not penetrate very deeply people are more likely to try and deafen out surrounding noise by increasing the listening volume [says who? You can't just say things.].

Continued

Figure 11.3. **Student Articles Edited Using Track Changes** (Continued)

Example 2

This story has no attribution. What are your sources? It also reads like notes. You have a bunch of facts, but no story. Why are writing about this? Do you know someone who suffers from Osgood-Schlatter? If so, talk to them.

Osgood-Schlatter Disease

Osgood-Schlatter disease is among the most common causes of knee pain in kids. However, there is still no immediate cure. **[source?]**

[this history is not essential to the story and may get cut in the final version.] Osgood-Schlatter was first described by Paget in 1891, but was not officially published or named until 1903. In 1903, Osgood-Schlatter was named after two men named Robert Osgood and Carl Schlatter. Osgood-Schlatter is not the only knee injury there is. There is also one very similar, which is caused by fractures of the tibial tubercle. They are hard to tell apart because there is no precise definition of Osgood-Schlatter.

[what are your sources?] One study **[by who? When was it published? Where do they work?]** of 389 adolescents in the U.S., 21% reported Osgood-Schlatter were actively involved in sports as opposed to the 4.5% not involved. In Finland, there was another study involving athletes. The study showed that the disease affected 13% of athletes. Also, there was an MRI study **[by who, when did they do it, where are they located?]** of 20 patients with Osgood-Schlatter. The MRI showed that their patellar tendon **[what is the patellar tendon?]** was attached

Continued

Figure 11.3. **Student Articles Edited Using Track Changes** *(Continued)*

---

more **[what does this next part mean? Avoid jargon, translate it in everyday English.]** proximally and in a broader area to the tibia. These studies show that the disease Osgood-Schlatter is not very common, but it is heard of.

Osgood-Schlatter disease occurs more often in boys than girls. This is due to the fact that more boys participate in sports and are more physically active than girls. Even though it's not as common in girls, they can still be affected. Most girls that are affected are between the ages of 10 and 11. However some still get it between ages 8 and 15. Boys on the other hand usually are diagnosed with Osgood-Schlatter between ages 13 and 14, but their ages can range from 12 to 15. These are the most common age ranges because Osgood-Schlatter is normally seen soon after a rapid growth spurt **[why?].**

CHAPTER 12

BEYOND WORDS

INSERT HERE

Most SciJourn teachers begin by having their students write an article. Given all the challenges facing first-time users of this project, that is probably a wise choice. In fact, we didn't even know if we should include a chapter on media other than writing, since doing justice to the topic of multiliteracies would probably result in another book.

Yet, it is important for readers to understand that the approach we describe in the previous pages can easily be adapted to various technologies, including video, podcasts, and graphics. All of the SciJourn standards still apply, but creating a different product requires some additional skills. But just as reading and analyzing text is an important aspect of science literacy, so, too, is learning to read and analyze broadcast media.

For reluctant writers or tech-savvy teens, engaging in alternative approaches to creating stories is an attractive approach. Creating podcast and videos can also provide opportunities for students to work in teams and learn the type of 21st-century skills that administrators and politicians advocate. However, the reality of school life often encroaches on a good idea. Acquiring the necessary technology, finding the time, and developing the expertise can be formidable

challenges. Thus, some of these approaches may be best suited to a small group, the afterschool program, or the informal science setting.

An Easy First Step: Seeing the World Through Science

Images are a great way to engage students, but become especially useful when dealing with struggling readers and writers. A student (or the teacher) might, for instance, choose a photo for its visual appeal and ask, "Where's the science?" Students might then brainstorm—either in pairs, small groups, or as a class—about questions that this image might evoke.

In a close-up of an eye, for instance, students might begin with the reportlike topics such as vision or eye color, but then be pushed for 500-word story ideas. Use our top-rated question: "Why does that interest you?"

In a situation like this, don't be afraid of topic drift. The topic of eyelashes for instance might lead to a question about their function and then onto "Are there people who don't have eyelashes?" which can lead to an internet search. And, yes, there are people without eyelashes. The medical condition is called alpoceia. Sufferers, even young people, can lose their hair, including their eyelashes.

This might lead to questions about a new product for thickening eyelashes. How does it work? Or it might lead to a discussion about disposable contact lenses. This is precisely the kind of discussion that takes place in the heads of scientifically literate people. These are the kinds of connections we want to help students make and the kinds of curiosities we want them to develop. Helping to make this "idea hopping" apparent to teens is a good thing, especially if the query is backed up with a bit of research.

Extended Photo Captions

Some students dread writing, especially the task of putting together 500 words about science. To ease one struggling reader into the project, we asked her to take photographs during a field trip to a butterfly house and write a 150- to 200-word caption explaining each of the images. Thus, the captions identified the species, explained what has happening, included some interesting facts about butterflies, and even had quotes from the resident entomologist. Another student used this captioning approach to write about the physics of cheerleading, using photographs to describe the forces demonstrated by the local team in action. The captions pointed out how cheerleaders use momentum, gravity, center of mass, distribution of weight, and other concepts to perform their routines. All the captions conformed to the SciJourn standards.

𝔉ront-𝔓age Science

Infographics

Rob Lamb, a chemistry teacher, has pioneered the idea of SciJourn infographics. Infographics are the visually rich charts made famous by *USA Today*. Unlike standard charts or graphs, infographics use a rich palette of imagery to present multiple dimensions of data. Icons, color, size, location, direction, images, and other elements can be tapped to present a wealth of data. *Good* magazine (*www.good.is*) regularly posts infographics on a diverse set of topics, including scientific and technological, that can serve as models. Creating infographics also helps students learn to read multimodal texts, a clear goal of people interested in literacy development and assessment.

One SciJourn teen used Photo-Shop to generate his infographic, but this required a sophisticated understanding of the software. By contrast, two teens in Rob's class cut and pasted different-size images of footballs, soccer balls, baseballs, and so on onto a poster board. The balls represented the numbers of teens injured in their respective sports each year, as reported by the Consumer Product Commission (Photograph 12.1). It was an easy and elegant solution.

Photograph 12.1

A "low-tech" infographic detailing the numbers of sport injuries to young people.

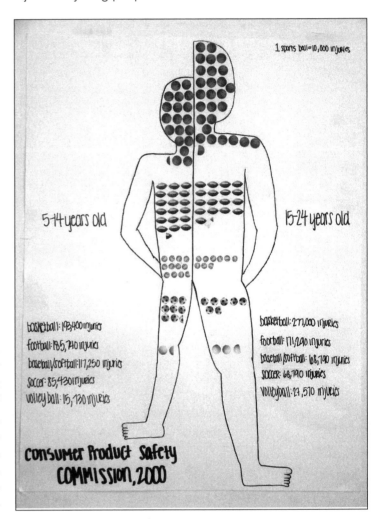

To create original infographics, Rob now uses the *iWork* applications *Numbers* for data and *Pages* for design, as well as the website *iconspedia.com* for imagery. Icons are clicked and dragged from iconspedia into Pages, and scaled for size. A textbox placed next to the icon provides explanation. For the detailed techie version of this lesson, see *Teach4SciJourn.org*.

As a warm-up, Rob has the class pull out their cell phones and determine how many apps, text messages, calls, music, videos, and photos they carry with them. (The data is shared with students lacking phones.) He then takes them

Figure 12.1. Classroom CO$_2$ Levels

The data clearly show that CO$_2$ levels rise when the room is occupied. Source: Monica Macheca.

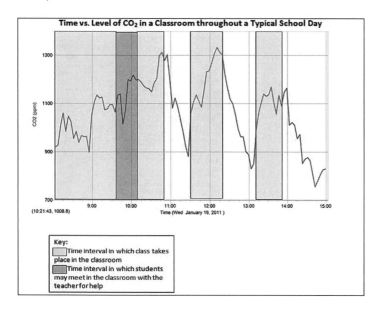

through a tutorial in which they create an infographic with the information gathered in the classroom. The next step is for the student or student team to generate a topic, collect appropriate data and create their own infographic.

A less technology intense version of the infographic is the captioned graph. Excel in Microsoft Office is an easy way to generate graphs of all sorts. One student, who had published an article in *SciJourner. org* about how CO$_2$ buildup in old buildings with poor ventilation can lead to sleepy students, decided to see if the effect was evident in her school. Working with her teacher, she launched a firsthand study of the CO$_2$ levels in her chemistry class (Figure 12.1). This type of data display is an ongoing area of research for the Sci-Journ team and we expect to publish more about it in the future.

Podcasts

For students who have trouble writing, but less trouble verbally articulating their ideas, podcasts present an appealing alternative form of presentation. It is still essential to have a script, but some adlibbing is fine. In our experience, it may take several "takes" until the whole podcast sounds right.

In terms of technology, digital recorders and computer-mounted microphones work well; the bigger problem has been to find a quiet space to record the story. If there is no sound studio, a closet works surprisingly well.

By keeping the recording short, the podcaster can avoid sound editing. Teens have tended to simply repeat the text until the whole story sounds right or right enough to them. This activity provides practice presenting in a very forgiving environment. However, if needed, the free application Audacity can be used to edit and create basic podcasts on most computers available in schools; the Mac application GarageBand has even more features for editing podcasts.

Ideas for sharing podcasts include posting them on a website, e-mailing them, or perhaps posting them on iTunes. One nice thing about a podcast is that it takes up considerably less memory than a video.

Audio-grabbing software can be handy when turning an interview into a story. One *SciJourner* reporter was fortunate to interview Ira Flatow, host of NPR's *Science Friday*. She turned the interview into a podcast by clipping out the sections of Flatow's answers she wanted, then recording a separate introduction to each section. She needed only about 250 words to stich together the various sections; the bulk of the podcast was taken up with Flatow's remarks. Because she has a good "radio voice," the end result was pretty close to professional. While students may feel that "making a podcast" or "making a movie" is a way to do less writing, in fact they often do more writing than they realize, and sometimes with less struggle, because the writing is embedded with other media work.

Video

As YouTube has shown, it is surprisingly easy to make movies. Unfortunately, as YouTube also demonstrates, it is frightfully easy to make bad movies. Lack of technology can be a real impediment to creating something good in this case.

Probably the simplest and most successful movie is a voiceover with still images, either gathered from the web (see section below on copyright) or created by the reporter. Obviously, this style of video still requires a script and some creative thought about what images support the content being discussed. A "Ken Burns" effect of moving in and out of the image can help keep the visual interesting.

Apple's iMovie or Microsoft's Moviemaker can be used effectively for this type of work, and they allow would-be directors to work with video and sound separately. Microsoft's Photostory can be used for photo-movies overlaid with audio commentary.

Educators interested in learning more about how to create photo-movies or photo and video movies might benefit from signing up for some additional training. Local workshops are often sponsored by media outlets, community organizations, and universities. Two particularly vibrant networks of educators dedicated to helping teens make videos are available. See, in particular, the work spearheaded by the Center for Digital Storytelling (*www.storycenter.org*) and the Digital Youth Network (*www.digitalyouthnetwork.org*).

Photograph 12.2

SciJourner reporter Ariel Stavri recording an interview for the website using a flip camera.

Photo: Alan Newman.

More complicated efforts involve creating a TV-type of news story, with on-the-scene reporting (Photograph 12.2). Our Saint Louis Science Center group has created several such movies, with mixed success. We have worked with simple flip cameras and, where possible, tripods. Audio has been a huge challenge, especially when covering a noisy event. Even with lapel microphones, it can be difficult to get clean sound. The omnidirectional microphones on most cheap video cameras collect every rattle in the room, making them worthless in these situations. A camera with a connection for an external microphone is thus a necessity. Inexpensive wireless lapel microphones can be found, but handheld microphones can also work.

We strongly suggest that news videos be created in teams. One person handles the video camera, another takes photographs that help augment the video sections, and the videographer or a third person interviews people. In some cases, the third person just handles audio by monitoring sound through headphones and the video person films and asks questions. The work is pooled and one member of the group stitches the video and photos into a coherent story and responds to feedback from the rest of the group.

It is extremely important that the team practices with the equipment before going out. Part of our own audio problems stem from the fact that over and over again, in the heat of the moment, the sound person plugs the jack into the wrong connector! Also, make sure the batteries are charged or fresh, and bring extras. There are no "do-overs" in this type of media.

Because most schools block YouTube, we mount our videos on *vimeo.com*, which is another free website. Vimeo does allow the user to place some controls on the video, so that a videographer can block others from stealing the file; there is even the option of making the video private. Once mounted on Vimeo, we can easily embed the file onto *SciJourner.org*.

Google Map

For those who venture outdoors, Google mapping is a nifty technology that can be used to present both observations and sourced information. Using longitude and latitude coordinates, it is possible to pinpoint a position on a Google map; tag it with a flag, which is called a waypoint; and then embed photos, video, and text into the waypoint. Students can even hyperlink within the waypoint to other sites. The result is a multimedia bonanza—a map or satellite view with markers at strategic points offering visual and written information.

Our first attempt at a Google map documented interesting plants at the Missouri Botanical Garden one day in August. We plan to update the map with images in different seasons. Other map enthusiasts have used this technology to document field trips, wildlife walks, geological formations and urban ecology.

To create the map, we put together a team. One person carried a GPS (the GPS is standard on many phones these days, but we had access to handheld units) and another carried a still camera, while a third person brought along a video camera. If you happen to have a camera that combines movie and still image functions, you may not even need much of a team! Someone takes notes or records oral notes on a digital recorder. The result? An interactive map in which photos, video, and written descriptions of data are linked to the GPS coordinates on a Google map.

To build the map, view the tutorial on Google's map section. This is fast changing technology, and it pays to check out the latest instructions. It is not too difficult to create these maps, especially if you limit your work to photos and text. Moreover, with some internet searching, you can learn how to obtain or create novel waypoint designs for maps, or upload or download maps to and from various devices. There is a huge and interesting subculture in GPS mapping. Geocaching—hiding and finding little trinkets using GPS coordinates—is an international phenomenon. Perhaps a local expert can be found to help you in your map activities.

Copyright

Along with going after students for plagiarism (see SLAP, Chapter 4), we harangue them regularly about copyright. Many of the photos students are happy to steal from the internet are copyrighted. If the image is not going public, then at least ask the author to give credit to the source. However, in a public forum, such as *SciJourner.org*, copyright means you can't use it without permission. And, for commercial sources, that permission means a fee.

Fortunately, especially in science, there are loads of copyright-free images. Many of the images found on government websites are yours for the taking. As taxpayers, we own the images produced by such fine organizations as CDC, NIH, FDA, NOAA, and EPA. In fact, some sites, such as NASA, maintain huge banks of downloadable images. All they ask is a credit line, which they richly deserve. Don't overlook international organizations, such as the World Bank or World Health Organization, either. Some nonprofits also are generous with their images.

Commercial organizations are often happy to supply corporate logos or product photos for free, with proper attribution. Look for the press or public information page on the company website; the images are located there, maybe within a press release. Our *SciJourner* story and subsequent movie on why iPads are popular—a good piece of technology reporting—made liberal use of Apple-supplied images.

Warning: Companies are very protective of their images; you can't modify them and use them to smear their product. They employ lots of expensive lawyers to keep you from doing that.

You can write to copyright holders and ask for permission to use their figures or photos for "one-time use," which means they still hold copyright. Obviously, they will get a credit and you should tell them that. This typically works with university professors, who care more about getting teens excited in

science than shaking them down for money. It will also work with nonprofits, such as the American Chemical Society or the American Geophysical Union, if it is defined as educational use. However, the response time from big organizations can be slow.

There is an interesting middle ground on the copyright issue, which has grown up with the web. Creative Commons (*creativecommons.org*) allows anyone to use an image, video, or music within certain sensible restrictions. Young artists have embraced the movement and Creative Commons sections can be found on such sites as *Wikipedia*, *Flickr*, and *SpinXpress*. Internet searches with the words Creative Commons will turn up more. In fact, *SciJourner.org* stories adhere to Creative Commons rules.

The Big Question

Is video worth all the work it entails? For some students, yes, but for many, probably not. It is our experience that students find the old technology most fulfilling—print versions of their stories they can easily share with friends and family.

In a few years, however, we may answer this question differently. The skills students come with change rapidly in our technology-rich world.

What we can say, however, is that the ability to read media productions will be greatly enhanced by a student's familiarity with the SciJourn standards. The buzz among teachers who use the methods suggested in this book is that their students are hearing and seeing the world of science communications differently.

AFTERWORD

A recent headline in our local newspaper, *The St Louis Post Dispatch*, read "Scientific Take on Teacher Review." The article went on to describe the state's decision to use standardized test data to determine teacher quality. Of course we had thoughts about the content of the article, opinions about the usefulness of student standardized test data as a measure of teacher effectiveness and concerns about the ramifications of such a move. But set those aside and just look at the headline. Specifically look at the word *scientific*.

It happens all the time, the invocation of this word and words and phrases like it. "Scientists have found…" "As science proves…" At times, it's not hard to see why one faction of the population has come to mistrust all things science, why some have gone so far as to imply that using scientific data in policy-making is somehow anti-American.

When the concept of science is used to create an aura of objectivity and, often, to shut down any debate, we are troubled. We are reminded of Jay Lemke's words on this topic. He suggests that science education helps create this problem "by successfully convincing most students that the 'experts,' who talk science, are 'smarter' than they are—and by failing, at the same time, to actually teach most students how to talk science. This encourages a rule of 'experts,' an alienation from science, a sense of inferiority, and ultimately a fear and hatred of this powerful 'unknown'" (Lemke 1990).

Only a small percentage of our students will go on to be scientists, but we are convinced that all need to be a part of scientific conversations. In Lemke's words, all need to be able to "talk science." As we have argued earlier, those who can engage in these debates don't always begin with all the necessary content knowledge. What they do begin with, though, is confidence. They are not scared by the word *scientific*. They are not shut down by the phrase "well, scientists have proven it." Instead, they ask questions: "Which scientists? When? How?"

The work of bringing science news into the classroom is messy. It's certainly difficult work for state education departments to monitor; it doesn't lend itself to easy assessment. So, why do teachers do it? They do it, we've found, because it gives them a way to help students see that science matters. When students realize this, they become the kind of people who ask questions, who can find answers, who are not afraid. They become the kind of people who call up their doctor, a local pharmacist, or even the EPA.

Teachers want to connect their curriculum to their students' lives, but they know that it takes more than just spicing up their current content. As one of the SciJourn teachers said, "I can't just 'wow' students with science." Helping young people understand that science matters takes inviting this messiness into the classroom and giving students some control over the science they want to learn.

The topics of student articles are outside any curriculum we've ever seen—topics that include rare diseases, teenage concerns, and local issues. A teacher would have a hard time predicting many of the topics that have emerged. A state education department certainly never could.

In this same vein, we know that we have only started the conversation about science news in schools, that this is not the final word. Just as teachers can't predict all the directions students will go, we are also certain that we can't predict all the ways teachers may use science news. We invite innovation and creativity and we look forward to seeing next moves. We know good teachers take ownership of what they bring to the classroom. As

teachers ourselves, we understand the power of a good idea in the hands of an excellent practitioner.

Just as we were putting the finishing touches on this book, we came across a blog post by one of our favorite educational theorists, James Paul Gee. In it he takes aim at what he calls a "mindless progressivism," a theory that children learn best by participation and immersion in interest-driven activities. In it he notes that "people can participate in an interest-driven group and still gain few of the higher-value skills that participation in the group leads others to attain. That is why an emphasis on production is important. Learning to produce the knowledge or outcomes an interest-driven group is devoted to leads to higher-order and meta-level thinking skills" (Gee 2011).

Over the course of this project we have often asked ourselves, how important is it that the students write? Might they not learn the skills we identify in our standards more efficiently if they didn't take the time to write and revise and revise yet again? How important is it to publish or at least to have an opportunity to publish? For us, Gee provides that answer when he says that, "If only a few are producers and most are consumers, then a group is divided into a small number of 'priests' (insiders with 'special' knowledge and skills) and the 'laity' (followers who use language, knowledge, and tools they do not understand deeply and cannot transform for specific contexts of use)."

Gee proposes instead—and we surely believe he is right—that teachers are best served by recognizing and signing onto what he calls "situated learning." This sort of learning requires well-designed learning environments that feature the following:

1. Multiple routes to full and central participation for all members of a group, a group organized around an interest and a passion to which the interest might lead.

2. Multiple routes to everyone learning to produce the knowledge, dispositions, skills, and tools necessary to sustain,

extend, and transform the interest and the passion.

3. Motivation and the desire to explore. The interest must then be channeled into a passion so that learners persist towards mastery via a great many hours of practice. Otherwise learners need to find another interest that will lead to a passion.

4. A curriculum designed so that learners are immersed in well-structured, well-designed, well-mentored and well-ordered problem solving inside experiences where goals are clear and action of some sort must be taken.

5. Copious feedback. Lots of data on multiple variables across time is collected and used to assess learners; assess their growth and development over time; and assess, compare, and contrast (for both learners and stakeholders) different possible trajectories to mastery, including ones that lead to innovation and creativity.

6. Learning and assessment that are so tightly integrated that finishing a level of learning is a guarantee of mastery at that level, a guarantee that learners can solve problems and not just retain facts (but use facts as tools for problem solving), and a guarantee that learners are well prepared for future learning.

7. Mastery by all learners of one or more specialties at a deep level, so that they are able to teach that specialty to others, and able to learn new things from others when needed.

8. An ability on the part of each learner to pool their specialty with other people's different specialties, and integrate their specialty with other people's specialties by seeing the "big picture" so as to

be able to solve problems that no one specialty can solve.

9. Mentoring by teachers and peers at various levels, as well as by the presence of smart tools and well-designed problem-solving environments (both real and virtual). All learners must learn to mentor.

10. A learning environment (think of teachers as creators of learning environments) that meets all the above conditions and that assesses people's learning in an adaptive and contextually responsive way.

11. Direct instruction and texts that are offered "just in time" (when learners can put them to use and see what they really mean) or "on demand" (when learners feel a need for large amounts of instruction or text in their trajectory of problem solving).

12. A situation that enables failure to be used as a learning device, so that the price of failure is, at least initially, kept so low that all learners are encouraged to explore, take risks, and try different learning styles.

13. A situation in which learners are shown through modeling and made well aware of adult or professional norms for the skills and dispositions they are developing and held to high standards based on those norms in ways that make clear that every learner can reach those norms should they choose to put in the time and effort.

14. An environment in which learners come to see and be able to use the relationships and connections among different types of skills and knowledge, often "stored" in different people, as well as to understand the larger social, environmental, and cultural implications of any proposed solution to a problem.

15. A situation in which learners can integrate and see the connections among science, mathematics, social science, the humanities, ethics, and civic participation. In today's world this often means seeing how the same social and digital tools can be used for different types of discovery and interventions in the world across the arts, sciences, and humanities.

16. An environment in which learners are well prepared to learn new things, make good choices, and be able to create good learning environments for themselves and others across a lifetime of learning.

17. A designed environment in which learners become well prepared to be active, thoughtful, engaged members of the public sphere (this is the ultimate purpose of "public" education), which means an allegiance to argument and evidence over ideology and force and the ability to take and engage with multiple perspectives based on people's diverse life experiences defined not just in terms of race, class, and gender, but also in terms of the myriad of differences that constitutes the uniqueness of each person and the multitude of different social and cultural allegiances all of us have.

It is our hope to these standards that the Sci-Journ project, at least in theory if not always in practice, aspires. Our success will be determined not by the advice proffered in these pages, but rather by the teachers who take seriously the goals Gee has so elegantly enumerated. Please use the ideas we propose in this volume to make it happen.

References

Gee, J. March 9, 2011. *Beyond mindless progressivism.* Available online at *www. jamespaulgee.com/node/51*

Lemke, J. 1990. *Talking science.* Westport, CT: Ablex.

Page numbers printed in **boldface** type refer to figures.

A

Abbott, Colleen, 121, 124
Academy of Sciences, 106
Accuracy. See Factual accuracy
Achievement, 28
Advertising, 48, 49
Affiliations, 125, 130
After-school programs, 25, 177
Alternative approaches to creating science
 news stories, 31, 177–184
 big question and, 184
 copyright and, 183–184
 extended photo captions, 178
 Google mapping, 182–183
 infographics, 178–180, **179, 180**
 podcasts, 180–181
 seeing the world through science, 178
 video, **181,** 181–182
American Cancer Society, 94
American Chemical Society, 106, 183
American Geophysical Union, 99, 106, 183
American Heart Association, 8
American Lung Association, 8
Andrew-Vaughan, Sarah, 139
Angle to story, 17, 20, 42, 67–86. See also
 Topics for stories
 checking accuracy of, 125–126, 131
Article status board, 135, **136**
Assessment(s), 2, 148–153
 assigning grades, 170–171
 Calibrated Peer Review, 143, 148–150, 152–
  153
 formative, 140
 learning and, 186
 rubrics and, 3, 39–40, 53, 146
 Science Article Filtering Instrument, 143, 149,
  150–153, **151,** 160
Attention to detail, 123. See also Factual
 accuracy
Attribution(s), 17, 19–20, 44–45, 65, 89, 98–99,
 89, 98–99
 checking accuracy of, 126
 importance of, 99
 incorporating into news articles, 99
 student problems with, 156
Audacity software, 180
Audio-grabbing software, 180
Authentic activities, 3–4, 11–12, 26, 27, 28, 30,
 40, 67, 107, 144, 149

B

Background knowledge, 92–94
Baker, Patricia, 77
Balance in story, 18–19
Berendzen Sam, **63**
Breaking the rules, 21

C

Calibrated Peer Review (CPR), 143, 148–150,
 152–153
CDC (Centers for Disease Control and
 Prevention), 8, 44, 45, 92, 94, 183
Center for Digital Storytelling, 181
Centers for Disease Control and Prevention
 (CDC), 8, 44, 45, 92, 94, 183
Changes in scientific information, 1–2
Classroom discussions, 27–28
Cognitive science, xi, xii
Collaborative teaching, 24
Comments in story, 18–19
Community and distributed knowledge, 5, 12, 27
Computer access, 28–29
Conceptualization/background of story, 18
Conferencing
 as prewriting activity, 140
 in response to writing, 160–162
Contacts for interviews, **105,** 105–106
Contextualizing information, 41, 45–46, 117–120,
 131, 142, 156
Cooperative learning, 2, 27, 32
Copyright, 183–184
CPR (Calibrated Peer Review), 143, 148–150,
 152–153
Creative Commons rules, 183–184
Credibility, 15, 17, 43–44
 assessment of, 88–91
  in the classroom, 88–90
  lesson plans for, 100–101

on the web, 91, 94, 100–101
fact-checking for maintenance of, 123
of press releases, 62
read-alouds/think-alouds about, 62–64
of scientific theories, 121
of story ideas, 81
Critical thinking, 3, 11, 12, 25, 91, 156
Curriculum, 2, 11–12, 185, 186
article topics for reinforcement of, 75
connecting read-aloud/think-alouds to, 64

D

Daily news briefings, 64
Davidson, Rose, 144
Details of story, 18
attention to, 123 (See also Factual accuracy)
specificity and precision of, 20
Digital Youth Network, 181
Diigo, 98, 101, 135, 140, 145, 146
Direct quotes from interviewees, 111, 113
Discovery Channel, 55
Discovery Research K–12 program, xi
Drafting, 134, 144–148. See also Writing
edit-alouds of, 146–148, **147, 165–166**
revising and, 48, 53, 134, 155–172
where to begin, 145–146
Dropbox.com, 135

E

Edit-alouds, 146–148, **147, 165–166**
Editor/editing, 15, 22, 67–68, 76, 118, 125
evolution of, 171–172
of podcasts, 180
SciJourner editing hierarchy, **163–164**
teachers and, 167–168
Editorials, 21
vs. reporting, 20
stereotyping and, 49
topics for, 85
Elevator speeches, **141**
E-mail, 161, 171
interviews by, 109, 110, 111, 138
schools blocking access to, 29
sharing podcasts by, 180
Emig, Janet, 145
End of story, 19
English teachers, 11, 14, 24, 124, 135, 167
Environmental Defense Fund, 96

Environmental Protection Agency (EPA), 94, 96,
104, 183, 185
Ethical issues
choice of topics, 31–32
copyright, 183–184
plagiarism, 48–49, 62, 71, 99, 124, 129, 139,
151, 169, 183
Expertise
of information sources, 12, 13–14
reading outside of one's area of, 40
teachers working outside area of, 29
Experts, xi, 4, 5, 7, 12
affiliations of, 125, 130
attribution to, 44
interviewing of, 29, **50,** 104, 106–107, 113
invitations to, 25, 66
locating, 105–106, 183
read-alouds/think-alouds by, 66
recognition of, 43, **50,** 83

F

Factual accuracy, **41,** 46–48, 117–118, 120–126
assertions, evidence and, 121
fact-checking, 123–126
attribution, 126
information, 124, 129
language, 124–125, 130
missing material, 126, 131
name, affiliation, and organization, 125,
130–131
problem or angle, 125–126, 131
learning to be accurate, 122–123
practicing fact-checking, 127
paragraphs for, **130**
Problems/Solutions chart for, 127, **127**
science literacy and, 121
of scientific theories, 121
striving for, 129–131
when mistakes are published, 129
when students are wrong, 127–129
Farrar, Cathy, **93**
Feedback/response to writing, 155–172
"actionable" written feedback, 168
different kinds of, 157–160
have appropriate fun with the "small" story,
159–160
play to the personal, 158–159
state a problem, 159

surprise the reader, 159
evolution of an editor, 171–172
responding to papers in writing, 162–167
responding to the whole class, 162
SciJourner editing hierarchy for, **163–164**
teacher guidelines for managing stack of
papers, 168–171
through conferencing, 160–162
using Track Changes, 171, **173–176**
when to start over, 157
for the whole class, 162
in writing, 162–167
5Ws in news articles, 21, 40, 46, 66, 144
Five-paragraph essay, 3, 4, 15, 29–30, 68, 142,
144, 162
Flatow, Ira, 180
Fleischer, Cathy, 139
Flickr, 184

G
Gaither, Linda, 77
GarageBand software, 180
Gee, James Paul, 186
Geocaching, 183
Goodin, Andrew, 146
Google mapping, 182–183
Google searches, 94, **95,** 100
GPS mapping, 182–183
Grading, 170–171
rubrics for, 3, 39–40, 53, 146
Grammar errors, 167
Graphic organizers, 143
Graphs, 66
captioned, 179–180, **180**
infographics, 178–180, **179**
Grasser, Beth, 152
Greenpeace, 96
Group interviews, 107, 108
Group projects, 32, 140

H
Handouts, 143
Health literacy, 6–8
Hughes-Watson, Pamela, 145

I
"I" stories, 21
iMovie software, 181

Infographics, 178–180, **179, 180**
Information
checking accuracy of, 124 (See also Factual
accuracy)
contextualizing of, 41, 45–46, 117–120, 131,
142
missing, 126, 127, 131
student problems with recognizing what is
important, 156
Interests of students, 27, 28, 72–74
Internet, 5, 11
access to, 28–29
assessing credibility of sources on, 91, 94
lesson plans for, 100–101
website wall, 101
guideposts approach to searching, 91
modeling a good search on, 91
school-blocked sites on, 29, 91, 182
searches and sources on, 94–98
search star, **97**
student guide to finding useful websites,
95–97
Interviews, 104–113
conducting, 110–112
deciding what to use from, 113
developing questions for, 107, 109–110
with family members, 107
first steps for, 106–107
group, 107, 108
kinds of, 107–108
lesson plan on who to interview, **109**
locating contacts for, **105,** 105–106
obtaining direct quotes during, 111, 113
by phone or e-mail, 109, 110, 111, 138
planning tools for, 112
preparation for, 108–110
protecting sources of, 112–113
recording or taking notes during, 111
role-plays of, 106–107
techniques for, 112
time allotted for, 111
iTunes, 180

J
Johnson, Damonte, **105**
Johnson, Hannah, 145
Journalism teachers, 24, 29, 124, 135
Jovanovic, Jennifer, **105**

K

KWL chart, 32

L

Lamb, Rob, 178–179
Language, checking accuracy of, 124–125, 130
Leading questions, 143
Learning
   assessment and, 186
   cooperative, 2, 27, 32
   hands-on/minds-on, 87
   keeping track of, 32
   scaffolded, xii, 3, 27, 40, 144
   "situated," 186
   well-designed environments for, 186–187
Learning logs, 32
Learning sciences research, xi, xii
Lede, **16,** 17, 138–139, 145
Lemke, Jay, 4, 185
Lessons, 9, 24–26
Local science events, 25
Lying, 48, 49

M

Mentor texts, 138–139, 146
Mini-lessons, 24
Misconceptions of students, 92, 127–129
Missing material, 126, 127, 131
Modeling
   read-aloud/think-aloud, 25, 26, 55–66
   search-aloud/think-aloud, 91, 100
Motivation, 27–28, 186
Moviemaker software, 181

N

Names, checking accuracy of, 125, 130
National Aeronautics and Space Administration
   (NASA)*,* 96, 183
National Cancer Institute, 99
National Institutes of Health (NIH), 8, 43, 49, 92,
   96, 123, 183
National Oceanographic and Atmospheric
   Association (NOAA), 96, 183
National Press Club, 108
National Research Council (NRC), 2, 3
National Science Education Standards, 2
   compared with SciJourn standards, **37–38**
National Science Foundation (NSF), xi, 4, 12, 88

National Society of Black Engineers, 106
Newman, Alan, 4
Newton's Laws, 121
NIH (National Institutes of Health), 8, 43, 49, 92,
   96, 123, 183
NOAA (National Oceanographic and
   Atmospheric Association), 96, 183
NPR, 110, 180
NRC (National Research Council), 2, 3
NSF (National Science Foundation), xi, 4, 12, 88
Numbers, 20
   checking accuracy of, 124
Nutgraf, **16,** 17, 66

O

Organization names, checking accuracy of, 125,
   130

P

Personally meaningful story ideas, 72–74
Photographs and captions, 65, 178
   copyrighted, 183–184
   extended captions, 178
Photo-movies, 181
PhotoShop, 179
Photostory software, 181
PIOs (public information officers), 85–86, 106
Pitching ideas for stories, 40, 67–68, 71, 72,
   75–79, 88, 96, 127–128, 140, 156, 162.
   See also Topics for stories
   basic pitch, 76
   power pitch, 76–77
   responses to, 77–85
   speed pitch, 76, 77
   strategies for, 76–77
   warm-up pitch, 76
Plagiarism, 48–49, 62, 71, 99, 124, 129, 139,
   151, 169, 183
Planned Parenthood, 8
Podcasts, 180–181
PowerPoint presentations, 31, 76–77, 134, 140,
   162
Precision in story, 20. See also Factual accuracy
Press conferences, 107
Press offices, 88
Press pass, **111**
Press releases, 49, 88
   vs. news articles for read-alouds/think-alouds,

61–63
Prewriting, 134, 135, 138–140. See also Writing
    assessing student understanding of research,
        139–140, **141**
    conferencing for, 140
    definition of, 138
    importance of mentor texts for, 138–139
    read-alouds and, 139
Problem solving, 28
Problems/Solutions chart, 127, **127**
Professional development, 12
    Teach4SciJourn.org, 8, 13, 19, 25, 30, 53, 76,
        91, **93,** 98, 99, 100, 108, 110, 112, 127,
        135, 139, 150, 167, 179
Professional societies, 106
Public information officers (PIOs), 85–86, 106
Publishing, 30–31, 134, 186

R
Rapport with students, 57
Read-alouds/think-alouds (RATAs), 25, 26,
        55–66
    annotated SciJourner.org article for, 59, **60–61**
    benefits of, 56, 57
    after completion of, 61
    creating template for, 58
    evaluating teaching efforts related to, 59
    for fact-checking, 127
    how-to guidelines for, 58–59
    lesson ideas for, 63–66
    making connections with, 56–58
    press releases vs. news articles for, 61–63
    prewriting and, 139
    by scientists, 66
    sources of materials for, 55, **59**
    strategies for, 57–58
    when to use, 59
Reading skills, 2–3, 11, 56, 57
Real-world problem solving, 28
Reporting, 5–6, 85–86, 103–115. See also
        Science journalism
    vs. editorializing, 20
    interviews for, 104–113
    surveys for, 103, 114–115
Research support for story ideas, 70–72
Respecting students' ideas, 78
Responding to ideas for stories, 77–85
Revision(s), 48, 53, 134, 155–172

"actionable" written feedback for, 168
different kinds of feedback for, 157–160
    have appropriate fun with the "small" story,
        159–160
    play to the personal, 158–159
    state a problem, 159
    surprise the reader, 159
evolution of an editor, 171–172
importance of, 155–156
number of, 156
process activities for, 156–157
providing feedback through conferencing,
    160–162
responding to papers in writing, 162–167
responding to the whole class, 162
SciJourner editing hierarchy for, **163–164**
student problem areas and, 156
teacher guidelines for managing stack of
    papers, 168–171
using Track Changes, 171, **173–176**
vs. when to start over, 157
Roth, Wolf-Michael, 5
Rubrics, 3, 39–40, 53, 146
Ruby, Mike, 56–57, 63

S
SAFI (Science Article Filtering Instrument), 143,
        149, 150–153, **151,** 160
Saint Louis Science Center, 25, 105, 181
Scaffolding learning, xii, 3, 27, 40, 144
Science Article Filtering Instrument (SAFI), 143,
        149, 150–153, **151,** 160
Science fair, 25, 75, 85, 118
Science journalism, xii, 5–6, 14–22, 133–153.
        See also Writing
    editing and, 15, 22
    goal of, 14
    information sources for, 14
    interviews for, 104–113
    maintaining credibility in, 15, 17
    standards for, 35–53 (See also SciJourn
        standards)
    structure of journalistic articles, 15–19, **16,** 66,
        142, 143
    teacher modeling of, 25, 26, 55–66 (See also
        Read-alouds/think-alouds)
    teaching journalistic form, 142–144
        comparing school reports with news

articles, 142
dangers and benefits of focusing with "exercises," 143–144
handouts, leading questions, graphic organizers, 143
Science literacy, xi–xii, 1–9
authentic examples of, 4
as community attribute, 5
definition of, 2, 3, 36
factual accuracy and, 121
goal of, 5
health literacy and, 6–8
science journalism in promotion of, 5–6, 88
SciJourn standards related to goal of, 36, 40–41, **41**
self-confidence and, 4
topics for articles and, 70
writing skills and, 3–4
Science news stories, 14–22
alternative approaches to creation of, 31, 177–184
big question and, 184
copyright and, 183–184
extended photo captions, 178
Google mapping, 182–183
infographics, 178–180, **179, 180**
podcasts, 180–181
seeing the world through science, 178
video, **181,** 181–182
angle to, 17, 20, 42, 67–86 (See also Topics for stories)
attribution in, 17, 19–20
authenticity of, 30
breaking the rules in, 21
contextualizing information in, 41, 45–46, 117–120
evaluating quality of, 64
factual accuracy of, **41,** 46–48, 117–118, 120–126
5Ws in, 21, 40, 46, 66, 144
ideas for, 39–40
incorporating attributions into, 99
numbers and specificity in, 20
original reporting in, 103–115
pitching ideas for, 40, 67–68, 71, 72, 75–79, 88, 96, 127–128, 140, 156, 162
plagiarism of, 48–49, 62, 71, 99, 124, 129, 139, **151,** 169, 183

vs. press releases for read-alouds/think-alouds, 61–63
publishing of, 30–31
reporting vs. editorializing in, 20
revising of, 48, 53, 134, 155–172
sources of information for, 87–101
structure of, 15–19, **16,** 66, 142, 143
up-to-date, 21, 79
writing of, 22, 29–30, 133–153
feedback/response to, 155–172
by teachers, 33
Science news websites, 55, **59**
Science skills and processes, 26, 46, 64
Science teachers
articles written by, 33
collaborative teaching by, 24
editors and, 167–168
guide to article status board, **136**
keeping track of student learning, 32
modeling by, 25, 26, 25, 26, 55–66 (See also Read-alouds/think-alouds)
setting up a writing classroom, 133–135
student conferences with
as prewriting activity, 140
as response to writing, 160–162
student rapport with, 57
Teach4SciJourn.org for, 8, 13, 19, 25, 30, 53, 76, 91, **93,** 98, 99, 100, 108, 110, 112, 127, 135, 139, 150, 167, 179
working outside one's area of expertise, 29
writing feedback provided by, 155–172 (See also Feedback/response to writing)
Science textbooks, 12
Scientific inquiry methods, 46
SciJourn, xii, 6–9
alternatives to, 31, 177–184
fitting into curriculum, 23–27
five-paragraph essay model and, 29–30
getting started with, 33
goal of, 12
as group project, 32
keeping students motivated for, 27–28
keeping track of student learning in, 32
practical concerns about, 23–33
professional development for use of, 12
publishing in, 30–31
read-alouds/think-alouds in, 55–66
sequencing of, 32

technology needed for, 28–29
topics for, 2, 26, 28
uncomfortable topics and ethical issues,
        31–32, 48–49, 74–75
working outside one's area of expertise, 29
SciJourn Blitzes, 24
SciJourn standards, 35–53, **41,** 65, 134
annotated article showing, **52,** 53
as aspirational target, 35, 53
compared with National Science Education
        Standards, **37–38**
content standards and, 36
development of, 39–41
ethics and, 48–49
related to science literacy goals, 36, 40–41,
        41
as rubrics, 53
Standard I: elements of article, 42–43, 68, 70
Standard II: information from relevant,
        credible sources, 43–44
Standard III: use of multiple, credible,
        attributed sources, 44–45
Standard IV: contextualizing information,
        45–46, 117
Standard V: factual accuracy and important
        information, 46–48
student version of, 40, **50–51,** 143
use of, 49–53
SciJourner, 6, **7,** 12, 13–14
editing hierarchy, **163–164**
press pass for, **111**
writing standards for, 35–53 (See also
        SciJourn standards)
SciJourner.org, 6, 7, 12, **13**
annotated article for read-aloud/think-aloud,
        59, **60–61**
example of story published in, **69**
goal of publication in, 134
ideas for article topics on, 72
purpose of, 30
SciJourn.org
Teach4SciJourn.org, 8, 13, 19, 25, 30, 53, 76,
        91, **93,** 98, 99, 100, 108, 110, 112, 127,
        135, 139, 150, 167, 179
Scipio, Déanna, 106
Scoring rubrics, 3, 39–40, 53, 146
Search-aloud/think-aloud, 91, 100
Searches and sources on the web, 94–98

assessing credibility of, 91
online tools for keeping track of, 98
search star, **97**
student guide to finding useful websites,
        95–97
Seeing the world through science, 178
Shamos, Morris, 5
Sierra Club, 96
Singer, Nancy, 145
Size of story topics, 68–70, 78, 79–80
too big, 79–80
too vague, 80
SLAP ethics, 48, 129, 152, **163,** 183
Society for Professional Journalists, 48
Sources of information, 14, 87–101
assessing credibility of, 43–44, 88–91
        in the classroom, 88–90
        lesson plans for, 100–101
        on the web, 91, 94, 100–101
attribution to, 17, 19–20, 44–45, 65, 89, 98–99
for background knowledge, 92–94
carousel activity related to, 90
definition of, 88
ethical concerns about, 31–32
expertise of, 12, 13–14 (See also Experts)
generalization from, 44
for interviews, 104, 105–106, 112–113
keeping track of, 98
multiple, 89
        credible, and attributed, 17, 19–20, 44–45
number of, 88
protection of, 112–113
relevance of, 43–44
SciJourn standards for, **41,** 43–45
searches and sources on the web, 94–98
        search star, **97**
        student guide to finding useful websites,
                95–97
Speaker visits, 25, 66
Specificity in story, 20
SpinXpress, 184
Stakeholders with interest in research, 40, 41,
        43, 44, **50,** 71, 104, 142, 186
Statistics, checking accuracy of, 124
Stavri, Ariel, **181**
Stereotyping, 49
Structure of journalistic articles, 15–19, **16,** 66,
        142, 143

comments and balance, 18–19
conceptualization/background, 18
details, 18
lede, 17
nutgraf, 17
triangle's apex, 19
Student–teacher conferences
    as prewriting activity, 140
    as response to writing, 160–162
Successes, 28
Surveys, 70, 103, 114–115
    guidelines for, 114–115
    how to use data from, 114
    school policies on, 114–115

T
Talking Science, 4
Teach4SciJourn.org, 8, 13, 19, 25, 30, 53, 76,
        91, **93,** 98, 99, 100, 108, 110, 112, 127,
        135, 139, 150, 167, 179
Technology needs, 28–29
Topics for stories, 2, 26, 28, 67–86, 185
    approval of, 79
    assigning of, 67
    blocked internet sites and, 29
    checking accuracy of, 125–126, 131
    credibility of, 81
    for curricular reinforcement, 75
    double-duty double cross about, 84
    editorials, 85
    editor's questions about, 76
    idea board for, **73**
    ideas for, 39–40, 67–68
    of interest to others, 74
    journalists' vs. teachers' views of, 68
    looking for stories behind, 78
    personally meaningful, 72–74
    pitching ideas for, 40, 67–68, 71, 72, 75–79,
        88, 96, 127–128, 140, 156, 162
    recognizing good ideas for, 77–78
    rejection of, 78
    research support for, 70–72
    respecting ideas for, 78
    science literacy and choice of, 70
    SciJourn standard for, 42–43
    size of, 68–70, 78, 79–80
    student misconceptions and, 92, 127–129
    uncomfortable or troublesome, 31–32, 74–75

up-to-date, 21, 79
when idea is good but direction is not, 83–84
when student is clueless about, 81–83
when teacher does not understand hook for,
        84
Track Changes, 171, **173–176**
TV-type news stories, 181–182

U
Up-to-date news articles, 21, 79

V
Vague topics, 80
Video, **181,** 181–182
Vimeo, 182

W
WebMD, 82, 91, 100, 103, 124
Websites. See Internet
WHO (World Health Organization), 44, 99, 183
Wikipedia, 71, 91, **95–97,** 124, 126, 184
World Bank, 183
World Health Organization (WHO), 44, 99, 183
Writing, 3–4, 22, 39, 133–153. See also Science
        journalism
    alternatives to, 31, 177–184
    article status board for, 135, **136**
    assessing student understanding of research
            before, 139–140, **141**
    assessment of, 143, 148–153
        Calibrated Peer Review, 143, 148–150,
                152–153
        Science Article Filtering Instrument, 143,
                149, 150–153, **151**
    assignment ideas for, 133–134
    criteria for completion of writing project, 134
    drafting, 134, 144–148
        edit-alouds, 146–148, **147, 165–166**
        where to begin, 145–146
    early completion of, 137–138
    feedback/response to, 155–172
    of first article, 134
    5Ws in news articles, 21, 40, 46, 66, 144
    five-paragraph essay, 3, 4, 15, 29–30, 68,
            142, 144, 162
    formulaic, 11
    goals of writing project, 133
    in groups, 140

importance of, 186
outlines for, 140
pacing of, 137–138
plagiarism and, 48–49, 62, 71, 99, 124, 129, 139, **151,** 169, 183
prewriting stage of, 134, 135, 138–140
publication and, 134
revising and, 48, 53, 134, 155–172
setting up a writing classroom, 133–135
6+1 traits scoring rubric for, 39–40
size of topic for 500-word essay, 68–70
speed feedback for, 140
stages in process of, 138
standards for, 35–53, 134 (See also SciJourn standards)

storing files, 135
student–teacher conferences about, 140
stumbling blocks for, 137
teaching journalistic form for, 142–144
    comparing traditional school reports with news articles, 142
    dangers and benefits of focusing with "exercises," 143–144
    handouts, leading questions, graphic organizers, 143
    when to start over, 157
Writing Outside Your Comfort Zone, 139

Y
YouTube, 181, 182